Numerical Solutions of Three Classes of Nonlinear Parabolic Integro-Differential Equations

Numerical Solutions of Three Classes of Nonlinear Parabolic Integro-Differential Equations

Temur Jangveladze
Ilia Vekua Institute of Applied Mathematics of
Ivane Javakhishvili Tbilisi State University
Tbilisi, Georgia
&
Georgian Technical University
Tbilisi, Georgia

Zurab Kiguradze
Ilia Vekua Institute of Applied Mathematics of
Ivane Javakhishvili Tbilisi State University
Tbilisi, Georgia

Beny Neta
Naval Postgraduate School
Department of Applied Mathematics
Montercy, CA, U.S.A.

ELSEVIER

AMSTERDAM • BOSTON • HEIDELBERG • LONDON
NEW YORK • OXFORD • PARIS • SAN DIEGO
SAN FRANCISCO • SINGAPORE • SYDNEY • TOKYO
Academic Press is an imprint of Elsevier

Academic Press is an imprint of Elsevier
32 Jamestown Road, London NW1 7BY, UK
225 Wyman Street, Waltham, MA 02451, USA
525 B Street, Suite 1800, San Diego, CA 92101-4495, USA
The Boulevard, Langford Lane, Kidlington, Oxford OX5 1GB, UK

ISBN: 978-0-12-804628-9

British Library Cataloguing in Publication Data
A catalogue record for this book is available from the British Library

Library of Congress Cataloging-in-Publication Data
A catalog record for this book is available from the Library of Congress

For information on all Academic Press Publications
visit our website at http://store.elsevier.com/

Contents

PREFACE

This book is concerned with the numerical solutions of some classes of nonlinear integro-differential models. Some properties of the solutions for investigated equations are also given. Three types of nonlinear integro-differential models are considered. Algorithms of finding approximate solutions are constructed and investigated. Results of numerical experiments with graphical illustrations and their analysis are given. The book consists of four Chapters.

In the first Chapter three models (Model I, Model II and Model III, that will be detailed further) to be discussed are introduced and a brief history of integro-differential equations is given.

In the second Chapter, mathematical modeling of a process of penetration of an electromagnetic field into a substance by integro-differential models is described. A short description of the integro-differential equation that is a special model for one-dimensional heat flow in materials with memory is also given in the second Chapter. This model arises in the theory of one-dimensional viscoelasticity as well. This Chapter closes with some concluding remarks of the three investigated models.

The third Chapter is devoted to the numerical solution of the initial-boundary value problems for models stated in the previous Chapter. Semi-discrete schemes and finite-difference approximations, as well as finite elements are discussed. The mathematical substantiation of all these questions for initial-boundary value problems is given.

The questions of the realizations of algorithms investigated in the third Chapter are discussed in the fourth Chapter. Results of the many numerical experiments with graphical illustrations and their analysis are also given in this Chapter.

At the end of the book a list of the quoted literature and indexes are given. The list of references is not intended to be an exhaustive bibliography on the subject, but it is nevertheless detailed enough to enable further independent work.

Each Chapter is concluded with a detailed section, entitled "Comments and bibliographical notes," containing references to the principal results treated, as well as information on important topics related to, but sometimes not included in the body of the text.

The authors believe that the book will be useful to scientists working in the field of nonlinear integro-differential models. In the opinion of the authors, the material presented in the book is helpful for a wide range of readers engaged in mathematical physics, in problems of applied and numerical mathematics, and also MS and PhD students of the appropriate specializations.

Temur Jangveladze,

Zurab Kiguradze,

Beny Neta

ACKNOWLEDGMENTS

The first author thanks Fulbright Visiting Scholar Program for giving him the opportunity to visit U.S.A. and the Naval Postgraduate School in Monterey, CA, U.S.A. for hosting him during the nine months of his tenure in 2012-2013. The second author thanks Shota Rustaveli National Scientific Foundation of Republic of Georgia for giving him opportunity to visit U.S.A. and the Naval Postgraduate School in Monterey, CA, U.S.A. for hosting him during the four months of his tenure in 2013.

Abstract

This book is concerned with the numerical solutions of some classes of nonlinear integro-differential models. Some properties of the solutions of the corresponding initial-boundary value problems studied in the monograph equations are given. Three types of nonlinear integro-differential models are considered. Algorithms of finding approximate solutions are constructed and investigated. Results of numerical experiments with tables and graphical illustrations and their analysis are given. The book consists of four chapters. At the end of the book a list of the quoted literature and indexes are given. Each chapter is concluded with a detailed section, entitled "Comments and bibliographical notes," containing references to the principal results treated, as well as information on important topics related to, but sometimes not included in the body of the text.

Numerical Solutions of Three Classes of Nonlinear Parabolic Integro-Differential Equation.
http://dx.doi.org/10.1016/B978-0-12-804628-9.50008-9

Key words: Electromagnetic field penetration, Maxwell's equations, integro-differential models, existence and uniqueness, asymptotic behavior, semi-discrete and finite difference schemes, Galerkin's method, finite element approximation, error estimate, stability and convergence.

Chapter 1

Introduction

Abstract

The description of various kinds of integro-differential equations and a brief history of their origin and applications are given. The importance of investigations of integro-differential models is pointed out as well. Classification of integro-differential equation is given. The main attention is paid on parabolic type integro-differential models. In particular, three types of integro-differential equations are considered. Two of them are based on Maxwell's equations describing electromagnetic field penetration into a substance. The third one is obtained by simulation of heat flow. At the end of the chapter, as at the end of each chapter, the comments and bibliographical notes is given, which consists of description of references concerning to the topic considered.

Key words: Electromagnetic field penetration, Maxwell's system, heat flow equation, integro-differential models.

Numerical Solutions of Three Classes of Nonlinear Parabolic Integro-Differential Equation.
http://dx.doi.org/10.1016/B978-0-12-804628-9.50001-6

In mathematical modeling of applied tasks differential, integral, and integro-differential (I-D, for short) equations appear very often. There are numerous scientific works devoted to the investigation of differential equations. There is a vast literature in the field of integral and integro-differential models as well.

The differential equations are connecting unknown functions, their derivatives, and independent variables. On the other hand, integral equations contain the unknown functions under an integral as well.

The term integro-differential equation in the literature is used in the case when the equation contains unknown function together with its derivatives and when either unknown function, or its derivatives, or both appear under an integral.

Let us recall the general classification of integro-differential equations. If the equation contains derivatives of unknown function of one variable then the integro-differential equation is called ordinary integro-differential equation. The order of an equation is the same as the highest-order derivative of the unknown function in the equation.

The integro-differential equations often encountered in mathematics and physics contain derivatives of various variables; therefore, these equations are called integro-differential equations with partial derivatives or partial integro-differential equations.

In the applications very often there are integro-differential equations with partial derivatives and multiple integrals as well, for example, Boltzmann equation [66] and Kolmogorov-Feller equation [288].

Volterra is one of the founders of the theory of integral and integro-differential equations. His works, especially in the integral and integro-differential equations, are often cited till today. The classical book by Volterra [469] is widely quoted in the literature. In 1884 Volterra [465] began his research in the theory of integral equations devoted to distribution of an electrical charge on a spherical patch. This work led to the equation, which in the modern literature is called the integral equation of the first kind with symmetric kernel.

The work on the theory of elasticity became the beginning research of Volterra leading to the theory of partial integro-differential equations. In 1909 Volterra [466] has studied a particular type of such equations and has shown that this integro-differential equation is equivalent to a system consisting of three linear integral equations and a second order partial differential equations.

The first examples of integro-differential equations with partial derivatives investigated in the beginning of the twentieth century were in Schlesinger's works [417], [418], where the following equation is investigated:

$$\frac{\partial U(x,y)}{\partial x} = \int_a^b f(x,y,s)U(x,s)ds.$$

Numerous works in the beginning of the twentieth century were devoted to research of integro-differential equations of various kinds. The excellent bibliography in this case is given in the classical book by Volterra [469]. In addition, Kerimov [271], the editor of the Russian translation of this book, has updated (up to 1970s) the list of references on integral and integro-differential equations.

Let us describe some classes of mathematical models of second order promoting intensive research on partial integro-differential equations.

When we take into account hereditary phenomena, the questions of physics and mechanics lead to integro-differential equations. A hereditary phenomenon occurs in a system when the phenomenon does not depend only on the actual state of the system but on all the preceding states through which the system has passed; that is to say, it depends on the history of the system and may therefore be called hereditary.

One of the important representatives of an integro-differential equations of elliptic type is the following equation connected with the hereditary phenomenon [467]

$$\Delta U(x,t) + \int_0^t \sum_{i=1}^{3} \frac{\partial^2 U(x,\tau)}{\partial x_i^2} f_i(t,\tau)d\tau = 0,$$

where $x = (x_1, x_2, x_3)$, Δ is a classical three-dimensional Laplace operator

$$\Delta U(x,t) = \frac{\partial^2 U(x,t)}{\partial x_1^2} + \frac{\partial^2 U(x,t)}{\partial x_2^2} + \frac{\partial^2 U(x,t)}{\partial x_3^2},$$

and f_i are known functions of their arguments. Let us note that here and below everywhere instead of x, y, z we use x_1, x_2, x_3 for the designation of Cartesian coordinates.

The mathematical modeling of the vibrations of an elastic chord in the case of linear hereditary process gives a hyperbolic type integro-differential equation [467]

$$\frac{\partial^2 U(x,t)}{\partial t^2} = \frac{\partial^2 U(x,t)}{\partial x^2} + \int_0^t \frac{\partial^2 U(x,\tau)}{\partial x^2}\phi(t,\tau)d\tau, \qquad (1.1)$$

where ϕ is a known function of its arguments.

One of the most important representatives of integro-differential models is the following nonlinear equation describing string vibration obtained by Kirchhoff [286] in 1876

$$\frac{\partial^2 U(x,t)}{\partial t^2} - \left[\lambda + \frac{2}{\pi}\int_0^{\pi} \left(\frac{\partial U(x,t)}{\partial x}\right)^2 dx\right] \frac{\partial^2 U(x,t)}{\partial x^2} = 0, \quad (1.2)$$

where λ is a known constant. Many authors investigated equation (1.2) and its natural generalizations:

$$\frac{\partial^2 U(x,t)}{\partial t^2} - a\left[\int_0^\pi \left(\frac{\partial U(x,t)}{\partial x}\right)^2 dx\right] \frac{\partial^2 U(x,t)}{\partial x^2} = 0 \qquad (1.3)$$

and

$$\frac{\partial^2 U(x,t)}{\partial t^2} - a\left(\left\|A^{\frac{1}{2}} U(t)\right\|\right) AU(t) = 0, \qquad (1.4)$$

where $a(S) \geq a_0 = const > 0$ is a known function of its argument and A is a self-adjoint positive operator, i.e., $A = A^* > 0$. The norm used in (1.4) is the one defined on the range of the operator A.

In investigating (1.3) and (1.4) type models it is sufficient to mention the following publications: [9], [24], [36], [37], [55], [61], [65], [130], [309], [327], [328], [349], [355], [356], [357], [373], [380], [383], [390], [394], [395] though this list is not complete. Let us note that equations (1.3) and (1.4) are also called Kirch hoff equations. They, along with some similar equations, describe important physical processes, among which are linear and nonlinear dynamics of different dimensional bodies (see, for example, [36], [355], [373] and [471]).

In other questions connected with hereditary phenomena, one finds the integro-differential equations of a parabolic type, which were investigated by Evans [160]. These equations look like

$$\frac{\partial U(x,t)}{\partial t} - \frac{\partial^2 U(x,t)}{\partial x^2} + \int_{t_0}^t \frac{\partial^2 U(x,\tau)}{\partial x^2} A(t,\tau) d\tau = 0, \qquad (1.5)$$

where A is a known function of its arguments.

Integro-differential equations of parabolic type arise in the study of various problems in physics, chemistry, technology, economics, etc. One very important problem of applied type is generated by mathematical modeling of processes of electromagnetic field penetration into a substance and is described by the well-known Maxwell's equations [300]. In the works [187] and [188], complex system corresponding to nonlinear partial-differential equations was reduced to integro-differential form. If the coefficient of thermal heat capacity and electroconductivity of the substance are highly dependent on temperature, then the Maxwell's system can be rewritten in the following form (see [187] and [188]):

$$\begin{gathered}\frac{\partial W(x,t)}{\partial t} + \nabla \times [a(S(x,t))\nabla \times W(x,t)] = 0, \\ \nabla \cdot W(x,t) = 0,\end{gathered} \tag{1.6}$$

where

$$S(x,t) = \int_0^t |\nabla \times W(x,\tau)|^2 \, d\tau. \tag{1.7}$$

In system (1.6), $\nabla \times W$ and $\nabla \cdot W$ are the usual vector operators with respect to the variables $x = (x_1, x_2, x_3)$. Even one-dimensional scalar version of this model is very complicated and its investigation has been possible yet only for special cases. The one-dimensional scalar case of the model (1.6), (1.7) has the following form

$$\frac{\partial U(x,t)}{\partial t} = \frac{\partial}{\partial x}\left[a\left(\int_0^t \left| \frac{\partial U(x,\tau)}{\partial x} \right|^2 d\tau \right) \frac{\partial U(x,t)}{\partial x} \right], \tag{1.8}$$

where $a(S) \geq a_0 = const > 0$ is again a known function of its argument. Investigation of (1.6), (1.7), and (1.8) type models began in the works [138], [187], and [188]. Since then many scientific publications were devoted to the investigation of existence and uniqueness of their solutions under various types of initial and boundary conditions. In this respect, especially significant are the works [49], [50], [141], [146], [147], [219], [220], [238], [254], [256], [302], [303], [304], [305], [306], [307], [322], [333], [334], and reference therein. Authors of this book have also made contribution in this direction, for example, see [137], [138], [139], [140], [141], [145], [146], [147], [186], [187], [188], [219], [220], [223], [238], [247], [248], [253], [254], [256], [261], and reference therein.

Making certain physical assumptions in mathematical description of the above-mentioned process of penetration of electromagnetic field into a substance, Laptev [306] has constructed a new integro-differential model, which represents a generalization of the system introduced in [187] and [188]. Founded on Maxwell's system the following parabolic integro-differential model is obtained

$$\frac{\partial W(x,t)}{\partial t} = a\left(\int_0^t \int_\Omega |\nabla \times W(x,\tau)|^2 \, dx d\tau\right) \Delta W(x,t). \quad (1.9)$$

In the works [303], [305], and [306] for conditionally closed operators, an operator scheme is constructed. This scheme is used for (1.6), (1.7) type models to prove existence and uniqueness of solution of initial-boundary value problems. In the above-mentioned work [306] Laptev points out that for these so-called averaged integro-differential models (1.9), it is necessary to develop a different, special approach.

In systems (1.6), (1.7), and (1.9), $W = (W_1, W_2, W_3)$ denotes a vector, which is connected by a vector of a magnetic field $H = (H_1, H_2, H_3)$ and is the function of the variables (x_1, x_2, x_3, t), which we will shorten to (x, t).

The multidimensional scalar analogues of systems (1.6), (1.7), and (1.9) have the following forms:

$$\frac{\partial U(x,t)}{\partial t} = \nabla \cdot \left[a \left(\int_0^t |\nabla U(x,\tau)|^2 \, d\tau \right) \nabla U(x,t) \right] \tag{1.10}$$

and

$$\frac{\partial U(x,t)}{\partial t} = a \left(\int_0^t \int_\Omega |\nabla U(x,\tau)|^2 \, dx d\tau \right) \Delta U(x,t), \tag{1.11}$$

respectively.

In equations (1.10) and (1.11) we have $x = (x_1, \ldots, x_n)$ and the vector operator ∇U is given by

$$\nabla U = gradU = \left(\frac{\partial U(x,t)}{\partial x_1}, \ldots, \frac{\partial U(x,t)}{\partial x_n} \right) = (D_1 U, \ldots, D_n U) \, .$$

Some generalizations of the models (1.6)-(1.11) are given in the works [137], [139], [145], [219], [305], and [306]. One kind of these models has the forms:

$$\frac{\partial U(x,t)}{\partial t} - \sum_{i=1}^{n} D_i \left[a(S(x,t)) \, |\nabla U(x,t)|^{q-2} \, D_i U(x,t) \right] = f(x,t) \tag{1.12}$$

and

$$\frac{\partial U(x,t)}{\partial t} - a(S(t)) \sum_{i=1}^{n} D_i \left[|\nabla U(x,t)|^{q-2} \, D_i U(x,t) \right] = f(x,t), \tag{1.13}$$

where

$$S(x,t) = \int_0^t |\nabla U(x,\tau)|^q \, d\tau$$

and

$$S(t) = \int_0^t \int_\Omega |\nabla U(x,\tau)|^q \, d\tau.$$

The models of type (1.6)-(1.13) are complex and have been intensively studied by many authors. The existence and uniqueness of global solutions of initial-boundary value problems for equations and systems of type (1.6) - (1.13) were studied in [49], [50], [137], [138], [139], [140], [141], [145], [186], [187],[188], [219], [223], [238], [247], [248], [253], [261], [302], [303], [304], [305], [306], [307], [322], [333], [334], and in a number of other works as well.

The existence theorems that are proved in [137], [138], [139], [140], [141], [145], [187], [188], and [219] are based on a-priori estimates, modified Galerkin's method and compactness arguments as in [327], [328], [461], and [462] for nonlinear elliptic and parabolic equations.

For equation (1.8) with nonhomogeneous right-hand side and $a(S) = 1 + S$, or for equation (1.13) in the one-dimensional case with $q = 2$, such theorem for first initial-boundary value problem is proved in section 3.6.

The asymptotic behavior as $t \to \infty$ of the solutions of such models has been the object of intensive research in recent years, see [28], [29], [31], [32], [33], [34], [35], [139], [141], [145], [146], [147], [217], [218], [219], [220], [223], [227], [228], [229], [231], [232], [233], [234], [235], [236], [237], [238], [239], [240], [241], [242], [243], [245], [246], [247], [248], [251], [252], [253], [254], [256], [257], [261], [276], [277], [278], [280], [282], [283], and reference therein.

Another model considered in this book and studied by one of the authors of this monograph [376] is

$$\frac{\partial U(x,t)}{\partial t} = \int_0^t a(t-\tau)\frac{\partial}{\partial x}\sigma\left(\frac{\partial U(x,\tau)}{\partial x}\right)d\tau + f(x,t). \quad (1.14)$$

This equation serves as a very special model for one-dimensional heat flow in material with memory [196], [345]. It also arises in the theory of viscoelasticity [109], [344], [346]. It is an example in the general theory of equations of the form

$$\dot{U}(t) + \int_0^t a(t-\tau)g(U(\tau))d\tau = f(t) \quad (1.15)$$

on a Hilbert space H with g a nonlinear bounded operator [51]. The dot above U means time derivative.

MacCamy [345] and later Staffans [438] have discussed the existence, uniqueness, boundedness and asymptotic behavior of solutions of initial-boundary value problems for (1.14).

Many scientific publications are dedicated to the investigation of characteristics of (1.14) and (1.15) type equations (see, for example, [67], [70], [80], [110], [111], [162], [206], [433], and reference therein).

Numerous scientific works are devoted to construction of algorithms for the numerical solution of initial-boundary value problems for the above-stated models (see, for example, [20], [30], [32], [35], [80], [96], [125], [139], [145], [148], [220], [223], [227], [239], [241], [243], [244], [247], [248], [249], [250], [251], [252], [253], [255], [256], [257], [258], [261], [274], [276], [277], [279], [280], [281], [282], [283], [284], [318], [333], [376], [378], [423], [424], and [501]).

The purpose of the present monograph is the continuation of study and uniform description of results developed for the numerical solution of integro-differential models (1.6)-(1.14) and their generalizations. It should be noted that main attention is paid on theorems with proofs for corresponding one-dimensional equations.

The book consists of four chapters. In the next chapter mathematical modeling of a process of penetration of an electromagnetic field into a substance by integro-differential models and the conclusion of investigated models such as (1.6)-(1.13) is briefly given. We also give the mathematical model for heat conduction in materials with memory (1.14). Some features of this model are also given.

The third chapter is devoted to the numerical solution of the initial-boundary value problems for the models stated in the previous chapter. The semi-discrete schemes, finite difference approximations, Galerkin's method schemes, and finite elements are discussed. The mathematical substantiation of all these questions for various types of initial-boundary value problems is given.

Results of the various numerical experiments with graphical illustrations and their analyses are given in the fourth chapter.

At the end of the book a list of the quoted literature and indexes are given.

The list of references is not intended to be an exhaustive bibliography on the subject, but it is nevertheless detailed enough to enable further independent work.

Each chapter is concluded with a detailed section, entitled "Comments and bibliographical notes," containing references to the principal results treated, as well as information on important topics related to, but sometimes not included in the body of the text.

The authors believe that the book will be useful to the experts working in the field of nonlinear integro-differential models. In the opinion of the authors, the material presented in the book is helpful for a wide range of readers engaged in mathematical physics, in problems of applied and numerical mathematics, and also MS and PhD students of the appropriate specializations.

1.1 Comments and bibliographical notes

Mathematical models of many natural phenomena and processes can be described by the initial-boundary value problems posed for nonstationary partial differential and integro-differential equations and systems of such equations. Investigation and numerical solution of these problems are the actual sphere of mathematical physics and numerical analysis. One such partial integro-differential model describes the process of electromagnetic field penetration into a substance. In the quasi-stationary approximation, the corresponding system of Maxwell's partial differential equations can be rewritten in integro-differential form (1.6), (1.7) (see [187] and [188]).

Mathematical models describing electromagnetic processes and many relative phenomena are given in many scientific papers and books, see, for example, [92], [136], [300], [359], [360], [399], [419], [473], and references therein.

It is well known that electromagnetic field diffusion processes and many other important practical processes are simulated by Maxwell's systems of partial differential equations and Maxwell's-type systems as well (see, for example, [195], [197], [290], [297], [298], [300], [360], [426], [485], [486], [487], [489], [491], and references therein).

Many scientific works are devoted to investigation of various problems for Maxwell's and Maxwell's-type systems, see [4], [5], [56], [60], [101], [106], [120], [121], [124], [131], [139], [144], [145], [152], [158], [194], [195], [197], [208], [224], [225], [226], [247], [253], [275], [276], [290], [362], [365], [367], [397], [409], [427],

[452], [453], [454], [470], [475], [476], [477], [485], [486], [487], [489], [491], and references therein.

As we have already pointed out, by using Maxwell's system [300] for mathematical simulation of the process of electromagnetic field penetrating into a substance, new class of integro-differential models (1.6), (1.7), (1.8), and (1.10) arises (see [187] and [188]).

Integro-differential equations arise in many other practical processes as well, see, for example, [11], [21], [22], [36], [37], [55], [59], [74], [84], [119], [127], [128], [164], [185], [187], [188], [295], [298], [306], [310], [311], [312], [355], [356], [357], [363], [373], [380], [384], [405], [425], [459], [463], [465], [466], [467], [468], [469], [471], [484], and in a number of other works as well.

The motivation for studying integro-differential problems comes from the many physical models in such fields as electromagnetic wave propagation, heat transfer, nuclear reactor dynamics, and thermoelasticity. Besides the integro-differential equations arise in many spheres of human activity as well. For example, the second order fully nonlinear integro-differential equations are derived from the pricing problem of financial derivatives and optimal portfolio selection problem in a market [59]. In [84] nonlinear integro-differential equations that arise from stochastic control problems with purely jump Levy processes are considered.

Many problems of modern science and engineering can be described by partial integro-differential equations. Since quite a lot of these problems are time-dependent, most of them are evolution equations and especially nonlinear evolution parabolic equations, see [93], [177], [179], [180], [191], [216], [310], [311], [312], [322], [332], [345], [346], [358], [451], and references therein).

Many scientific works are devoted to investigation and numerical solution of parabolic integro-differential models, see [6], [18], [45], [53], [62], [63], [69], [74], [83], [86], [87], [93], [98], [103],

[108], [115], [118], [128], [137], [139], [145], [155], [156], [163], [175], [177], [181], [182], [191], [198], [247], [253], [263], [265], [266], [267], [276], [279], [281], [306], [308], [310], [311], [312], [313], [319], [320], [322], [323], [327], [353], [358], [370], [376], [378], [389], [393], [427], [430], [432], [433], [447], [448], [449], [450], [451], [456], [479], [483], [488], [492], [497], and references therein.

Study of the models of type (1.6), (1.7), and (1.8) has begun in the works [138], [187] and [188]. In these works, in particular, the theorems of existence of solution of the initial-boundary value problem (with first (Dirichlet) boundary conditions) for scalar equation with one-dimensional space variable are proved. Investigations of higher space dimensions for model (1.10) carried out initially in [137] and [140].

In [306] some generalization of the system of type (1.6), (1.7) was proposed. In particular, assuming that the temperature of the considered body depends on the time variable, but independent of the space coordinates, then the same process of penetration of a magnetic field into the material is simulated by the averaged (as the author of [306] has named it) integro-differential models (1.9) and (1.11).

Study of the models of type (1.9) and (1.11) has started in the works [217] and [219].

One must note that some works were devoted to the study of modeling of physical process of electromagnetic field penetration in the case of cylindrical conductors. In this case, the above-mentioned integro-differential model (1.6), (1.7), written in cylindrical coordinates, was given in [148]. The work [333] is devoted to the investigation of periodic problem for one-dimensional (1.8) type model in cylindrical coordinates.

Interest in the above-mentioned integro-differential (1.6) - (1.11) models is increasing. Some generalizations of (1.10) and (1.11) models, which have the forms (1.12) and (1.13) correspondingly, are given and studied in [219], [305], and in a number of other works as well. Equation (1.12) is investigated, for example, in [145], [219], [305], and [306]. Equation (1.13) was investigated in [145] and [219]. In the scientific literature some more general models have also appeared, see [137], [139], [145], [219], [305], [306], and [322].

Many different kinds of initial-boundary value problems with a variety of boundary and initial conditions are considered for the above-mentioned integro-differential equations. In the works [246], [280], [281] investigation and numerical approximation of problems with mixed boundary conditions, for (1.10) and (1.11) type one-dimensional scalar models, are studied.

Let us also note that first kind initial-boundary value problems with nonhomogeneous boundary condition on one side of lateral boundary are also considered and studied in many works. This type of the problem statement is dictated by mathematical simulation of the physical processes, see, for example, [148] and problem (2.64), (2.65) given in "Comments and bibliographical notes" section in Chapter 2.

The theorems proved for investigating the asymptotic behavior as time tends to infinity in some cases show the difference between stabilization character of solutions with homogeneous and nonhomogeneous boundary conditions of the first kind initial-boundary value problems. More precisely, in homogeneous case stabilization has an exponential character, whereas in nonhomogeneous case it has power-like form.

The works [49], [50] are also worth mentioning, where investigation of inverse problems for multidimensional models of (1.10) type is carried out.

Another integro-differential model studied in this monograph is (1.14). This model describes heat flow in material with memory [196], [345]. It also arises in the theory of viscoelasticity [109], [344], and [346].

As a rule we cannot find exact solutions of the considered nonlinear differential and integro-differential models. Therefore, particular attention should be paid to construction of numerical solutions and to their importance for integro-differential models. The first steps in this direction, for the models studied in this monograph, are made in the works [139], [220], [274], [376], and [378]. Now the research in this direction has intensified.

Let us note that the models being considered in this monograph have arisen from practical tasks. But they can be considered as models, generalizing known nonlinear parabolic equations, which are studied in many known scientific papers, books, and monographs, see [76], [133], [134], [169], [199], [298], [327], [332], [381], [385], [461], and references therein.

Many scientific researches are devoted to numerical solution of partial differential and nonlinear parabolic equations as well, see [23], [43], [104], [167], [171], [183], [193], [210], [213], [225], [226], [301], [327], [400], [404], [407], [411], [415], [435], [439], [440], [447], and the references listed in these papers and books.

As we have already mentioned the main part of integro-differential structures considered here has arisen from Maxwell's systems of partial differential equations. There are many scientific papers and books on the numerical solution of Maxwell's systems and Maxwell's-type systems as well, see [4], [5], [15], [26], [27], [54], [68], [139], [142], [145], [153], [212], [224], [247], [253], [260], [276], [314], [315], [316], [317], [336], [341], [369], [434], [442], [472], [480], [493], [498], and references therein. The results of these researches very often are used in construction and investigation of the numerical algorithms for the corresponding integro-differential models.

The detailed description of investigation and numerical solution of the above-mentioned models is given in Chapters 2-4. More complete references and comments are given in a section entitled "Comments and bibliographical notes" in each chapter.

Chapter 2

Mathematical Modeling

Abstract

The chapter consists of six sections. The chapter concerns mathematical modeling of the investigated in the monograph equations. Some mathematical features of these models are studies as well. In the first section general statement of diffusion process is given. Sections two and three are dedicated to a reduction of Maxwell's equations to the integro-differential models. Consequently, two types of integro-differential equations are obtained which are called Model I and Model II accordingly. Both models in different physical assumptions describe process of penetration of an electromagnetic field into a substance. In the third section Model III is considered, which represents a special model for one-dimensional heat flow in materials with memory. This model arises in the theory of one-dimensional viscoelasticity as well. Next two sections are devoted to some mathematical features of all three models above. The existence and uniqueness properties of the solutions as well as asymptotic behavior of solutions of the appropriate initial-boundary value problems are presented. This chapter closes with some concluding remarks and bibliographical overview of the three

Numerical Solutions of Three Classes of Nonlinear Parabolic Integro-Differential Equation.
http://dx.doi.org/10.1016/B978-0-12-804628-9.50002-8

investigated models.

Key words: Mathematical modeling, electromagnetic field penetration, Maxwell's equations, heat flow in materials with memory, one-dimensional viscoelasticity, integro-differential models, existence and uniqueness, asymptotic behavior.

2.1 Electromagnetic diffusion process

2.1.1 General statement of diffusion process

Let us consider the phenomena which occur in a substance placed in an external variable electromagnetic field.

Suppose the electromagnetic field and the currents satisfy quasi-stationary conditions, see, e.g., [300] and [419]. Maxwell's system of equations is given by:

$$-\frac{1}{c}\frac{\partial(\mu H)}{\partial t} = \nabla \times E, \tag{2.1}$$

$$\nabla \cdot (\mu H) = 0, \tag{2.2}$$

$$\frac{4\pi}{c}\sigma E = \nabla \times H, \tag{2.3}$$

where $E = (E_1, E_2, E_3)$ and $H = (H_1, H_2, H_3)$ are vectors of electrical and magnetic fields, respectively, μ is the magnetic permeability, σ is the conductivity of substance, and c is the speed of light in vacuum.

In (2.3), following an assumption of quasi-stationarity, the term proportional to $\partial E/\partial t$ is omitted and Ohm's law is used, relating the vector E to a vector of density of a current J by the following relation

$$J = \sigma E. \tag{2.4}$$

If at some moment of time the magnetic field is solenoidal, i.e., $\nabla \cdot (\mu H) = 0$, then (2.2) is always satisfied. From (2.1) we have

$$\frac{\partial}{\partial t}\nabla \cdot (\mu H) = 0.$$

We shall assume that the substance is isotropic with $\mu = 1$ and the coefficient of electroconductivity depends on temperature, i.e., $\sigma = \sigma(\theta)$. In applications, the form of σ is power-like, for example, for metals $\sigma \approx \theta^{-1}$, for homogeneous plasma $\sigma \approx \theta^{-3/2}$, and so on.

For definition of temperature it is necessary to use the equation of balance of heat. Let us make the assumption that the characteristic time of resistive diffusion is much less than that of heat transfer. Then, neglecting the effect of heat conductivity, the change in temperature is defined only by Joule's heating and taking into account (2.4), we have

$$c_v \frac{\partial \theta}{\partial t} = EJ = \sigma E^2, \tag{2.5}$$

where c_v is the specific heat capacity of the environment. Thus the factor of heat capacity can also depend on temperature (as a rule, in a power-like form).

Equations (2.1), (2.3), (2.5) form the closed system for definition of an electromagnetic field and temperature under an appropriate initial and boundary conditions.

The questions of existence and uniqueness of the solutions of linear differential problems ($\sigma = \sigma(x)$), in general enough statement, are considered in [136], [297], and in a number of other works. Thus, in [297] the movement of environment is supposed within the framework of magnetic-hydrodynamical approach. In these works the transition from classical statement to the generalized one is made. The requirement that the functions satisfy equation (2.2) and boundary conditions is replaced with the requirement of their belonging to special functional spaces. The problem for equations (2.1), (2.3) is formulated in the variational form, in particular, in terms of variational inequalities.

The study of system (2.1)-(2.3) can be made also on the basis of the equations which have been written down only for vectors E or H. As an example of similar research, see [504], in which the nonlinearity caused by dependence of the magnetic permeability μ on H is entered.

2.1.2 A reduction of system of nonlinear differential equations to the integro-differential model (Model I)

We assume that a massive body is placed in a variable magnetic field. It is necessary to describe the distribution of the field inside the body.

Inserting the resistance $\rho = 1/\sigma$ in (2.4) we shall express $E = \rho J$ and substitute in the penultimate equation (2.3), we have

$$E = \rho \frac{c}{4\pi} \nabla \times H.$$

Substituting this expression in (2.1), we receive

$$\frac{\partial H}{\partial t} + \frac{c^2}{4\pi} \nabla \times (\rho \nabla \times H) = 0. \tag{2.6}$$

A variable magnetic field penetrating into a substance induces a variable electrical field, which causes occurrence of currents. The currents result in heating of the substance and increase of its temperature, which influences its resistance. From reasonings given in [399], it follows that one order change of temperature results in change of resistance ρ on several orders, so at large fluctuations of temperature the dependence $\rho = \rho(\theta)$ must be taken into account. Last essential restriction, which will be made, is connected to the assumption that the change in body temperature under action of a current J submits to Joule's law

$$\rho_0 c_v \frac{\partial \theta}{\partial t} = \rho J^2. \tag{2.7}$$

Here ρ is density of the substance, ρ_0 is the density at time $t = 0$, and c_v (as was already mentioned in 2.1.1) is its specific heat capacity. Generally they also depend on temperature. Equation (2.7) does not take into account transfer of heat due to heat conductivity and radiation. Number of other physical effects are not considered. However, from the mathematical point of view system (2.6), (2.7) is complex enough.

Now we shall begin the reduction of system (2.6), (2.7) to a system of nonlinear integro-differential equations.

Let us rewrite equation (2.7) in the following form

$$\rho_0(\theta) \frac{c_v}{\rho(\theta)} \frac{\partial \theta}{\partial t} = J^2.$$

Introducing the function

$$S(\theta) = \int_{\theta_0}^{\theta} \rho_0(\xi) \frac{c_v}{\rho(\xi)} d\xi,$$

we have

$$\frac{\partial S}{\partial t} = J^2.$$

Let us assume that the process begins at the moment of time $t = 0$ in which the temperature θ_0 of the substance is constant.

Integrating this equation on a segment $[0, t]$, we shall find

$$S(\theta(x, t)) - S(\theta_0) = \int_0^t J^2 d\tau.$$

The functions ρ_0, ρ, and c_v are positive in physical sense; therefore, the function $S(\theta)$ is monotonically increasing. From here follows that the inverse function $\theta = \phi(S)$ is uniquely determined, connected with $S(\theta)$ by parity $\phi(S(\theta)) = \theta$. Thus, it is possible to write down

$$\theta(x,t) = \phi\left(\int_0^t J^2 d\tau\right).$$

From (2.3), (2.4) we have

$$J = \frac{c}{4\pi}\nabla \times H,$$

so,

$$\theta(x,t) = \phi\left(\int_0^t \left|\frac{c}{4\pi}\nabla \times H\right|^2 d\tau\right).$$

Substituting this expression in (2.6) as argument of function $\rho = \rho(\theta)$, we receive

$$\frac{\partial H}{\partial t} + \frac{c^2}{4\pi}\nabla \times \left[\rho\left(\phi\left(\int_0^t \left|\frac{c}{4\pi}\nabla \times H\right|^2 d\tau\right)\right)\nabla \times H\right] = 0,$$

$$\nabla \cdot H = 0.$$

Let us introduce the notations

$$a(S) = \frac{c^2}{4\pi}\rho(\phi(S)), \qquad W = \frac{c}{4\pi}H$$

and rewrite the system in the following form:

$$\frac{\partial W}{\partial t} + \nabla \times \left[a\left(\int_0^t |\nabla \times W|^2 d\tau\right)\nabla \times W\right] = 0, \tag{2.8}$$

$$\nabla \cdot W = 0.$$

If we consider a flat field of the form $W = (0, 0, U)$, where the function $U = U(x_1, x_2, t)$ depends on two spatial variables, then

$$\nabla \times W = \left(\frac{\partial U}{\partial x_2}, -\frac{\partial U}{\partial x_1}, 0 \right)$$

and (2.8) takes the form

$$\frac{\partial U}{\partial t} = \nabla \cdot \left[a \left(\int_0^t |\nabla U|^2 \, d\tau \right) \nabla U \right]. \tag{2.9}$$

Let us note again that equations such as (2.8), (2.9) for the first time have arisen in [187] and [188]. In these works, as well as in [137], [138], [139], [140], [141], [145], [186], [219], [223], [238], [247], [253], and [261] together with other questions, the uniqueness of the solutions of the initial-boundary value problems for equations (2.8) and (2.9) is given, under a rather general assumptions on the function $a = a(S)$.

Assume that ρ_0 and c_v are constants. Let us give examples of function $\rho(\theta)$, inducing $a(S)$, see, e.g., [145], [304], and [306].

If $\rho(\theta) = \theta^\alpha$, $\alpha > 1$, then

$$a(S) = C_1 (C_0 - S)^{\frac{\alpha}{\alpha - 1}},$$

where C_0 and C_1 are some positive constants. Thus, the power growth of the resistance $\rho(\theta)$ gives a coefficient determined only on a finite interval. Let us notice that the physical substances do not have such property.

If $\rho(\theta) = \theta$, then

$$a(S) = C_1 e^{C_0 S},$$

with positive constants C_0 and C_1, i.e., the linear growth of the function $\rho(\theta)$ gives exponential function $a(S)$. The linear growth of resistance with temperature is characteristic of metals.

If $\rho(\theta) = \theta^\alpha$, $0 < \alpha < 1$, then

$$a(S) = C_1 (C_0 + S)^{\frac{\alpha}{1-\alpha}},$$

with positive constants C_0 and C_1. Thus, under-linear growth of the function $\rho(\theta)$ gives sedate growth of the coefficient $a(S)$.

If $\rho(\theta) = \theta^\alpha$, $\alpha < 0$, then

$$a(S) = C_1 (C_0 + S)^{-\frac{\alpha}{1+\alpha}} > 0,$$

with positive constants C_0 and C_1, i.e., a decreasing function $\rho(\theta)$ gives decreasing functions $a(S)$. This is a general rule in case ρ_0 and c_v are constants. Suppose that function $\rho(\theta)$ is differentiable for $\theta \geq \theta_0$. Then according to the definition we have

$$\frac{da}{dS} = \frac{c^2}{4\pi} \frac{d\rho}{d\theta} \frac{d\theta}{dS} = \frac{c^2}{4\pi} \frac{d\rho/d\theta}{dS/d\theta} = \frac{c^2}{4\pi\rho c_v} \rho(\theta)\rho'(\theta).$$

From here it is obvious that the function $a(S)$ decreases or grows simultaneously with the function $\rho(\theta)$. Let us note that the decrease of resistance $\rho(\theta)$ with growth of temperature is characteristic of semiconductors in solid, gaseous, and plasma phases, for which, as it was already remarked in 2.1.1, the theoretical formula $\rho(\theta) = K\theta^{-3/2}$ is valid.

2.2 On the averaged Model II

We shall begin now a derivation of the averaged equation [306] describing again the process of penetration of an electromagnetic field into a substance.

Following [419] and as in section 2.1, we shall consider the system of the Maxwell's equations, describing the interaction of an electromagnetic field into a substance:

$$-\frac{1}{c}\frac{\partial H}{\partial t} = \nabla \times E, \tag{2.10}$$

$$\nabla \cdot H = 0, \tag{2.11}$$

$$\frac{4\pi}{c}J = \nabla \times H, \tag{2.12}$$

$$J = \sigma E. \tag{2.13}$$

Joule's law (2.7) is a localization of the law of allocation of heat [419]

$$dQ = \int_{\Omega} JE dx dt. \tag{2.14}$$

Here dQ is inflow of a thermal energy absorbed by a body in an electromagnetic field in time dt on a substance, occupying area $\Omega \subset \mathbb{R}^3$. Joule's law does not take into account the process of transfer of heat inside a body, which is fair if one to consider the temperature time-dependent but independent of the spatial coordinates. Thus, in this case it is possible to write that $\theta = \theta(t)$ and $dQ = mc_v\, d\theta$, where m is the weight of the substance. Therefore, (2.14) becomes

$$mc_v(\theta)\,\frac{d\theta}{dt} = \int_{\Omega} JE dx.$$

Using Ohm's law $E = \rho(\theta)J$ and repeating the process of exception, we shall receive the following analogue of system (2.8) in the same designations:

$$\frac{\partial W}{\partial t} + \nabla \times \left[a\left(\frac{1}{|\Omega|}\int_0^t \int_{\Omega} |\nabla \times W|^2\, dx d\tau\right) \nabla \times W\right] = 0,$$

$$\nabla \cdot W = 0.$$

Here $|\Omega|$ is the volume of the substance. Due to averaging the coefficient of this equation depends only on the variable t, and consequently the equation can be rewritten in the form

$$\frac{\partial W}{\partial t} = a\left(\int_0^t \int_\Omega |\nabla \times W|^2 \, dx d\tau\right) \Delta W. \tag{2.15}$$

Here we used the known vector identity

$$\nabla \times \nabla \times W = -\Delta W + \nabla(\nabla \cdot W).$$

For a flat field $W = (0, 0, U)$, where $U = U(x_1, x_2, t)$ is a function of two spatial variables, system (2.15) becomes

$$\frac{\partial U}{\partial t} = a\left(\int_0^t \int_\Omega |\nabla U|^2 \, dx d\tau\right) \Delta U. \tag{2.16}$$

Laptev [306] remarked that the research of the so-called averaged integro-differential models (2.15) and (2.16) requires independent theory than he applied for investigation of (2.8), (2.9), and such type equations.

2.3 Mathematical Model III

The classical linear theory of heat conduction for homogeneous and isotropic media is given by

$$\frac{\partial U}{\partial t} = k\frac{\partial^2 U}{\partial x^2},$$

where $U(x, t)$ is the absolute temperature and $k > 0$ is a constant called thermal diffusivity. This parabolic equation allows a thermal disturbance at any point in the medium to be instantly felt at other points, see [196]. Gurtin and Pipkin [196] have suggested a heat flow model based on a memory effect in the medium. The linear one-dimensional version of their theory assumes that the internal energy, $\epsilon(x,t)$, and the heat flux, $q(x,t)$, satisfy the following relations:

$$\epsilon(x,t) = bU(x,t) + \int_0^\infty B(\tau)U(x,t-\tau)d\tau, \tag{2.17}$$

$$q(x,t) = \int_0^\infty c(\tau)\frac{\partial U(x,t-\tau)}{\partial x}d\tau, \tag{2.18}$$

where b is the instantaneous specific heat, B is an energy relaxation function, and c is the thermal conductivity (see [345]). Thus ϵ is a functional of the history of the temperature and q is a functional of the history of the temperature gradient. MacCamy [345] treated a partially nonlinear version of the above model by replacing (2.18) by

$$q(x,t) = -\int_0^\infty K(\tau)\sigma\left(\frac{\partial U(x,t-\tau)}{\partial x}\right)d\tau. \tag{2.19}$$

Now assume that the material is at zero temperature and internal energy up to time $t = 0$. This is not restrictive since we can incorporate nonzero initial values with the forcing term. Incorporating (2.17), (2.19) with the balance of heat equation, we have

$$\begin{aligned} &b\frac{\partial U(x,t)}{\partial t} + \int_0^t B(t-\tau)\frac{\partial U(x,\tau)}{\partial \tau}d\tau = \\ &\int_0^t K(t-\tau)\frac{\partial}{\partial x}\sigma\left(\frac{\partial U(x,\tau)}{\partial x}\right)d\tau + r(x,t), \end{aligned} \tag{2.20}$$

where $r(x,t)$ is the heat source. Let $R(x,t)$ be the resolvant for B, i.e., the function U given by

$$U(x,t) = \frac{1}{b}\epsilon(x,t) + \int_0^t R(t-\tau)\epsilon(x,\tau)d\tau$$

solves the equation

$$\epsilon(x,t) = bU(x,t) + \int_0^t B(t-\tau)U(x,\tau)d\tau.$$

Now define $a(t)$ and $f(x,t)$ by

$$a(t) = \frac{1}{b}K(t) + \int_0^t R(t-\tau)K(\tau)d\tau,$$

$$f(x,t) = \frac{1}{b}r(x,t) + \int_0^t R(t-\tau)r(x,\tau)d\tau,$$

then (2.20) becomes

$$\frac{\partial U(x,t)}{\partial t} = \int_0^t a(t-\tau)\frac{\partial}{\partial x}\sigma\left(\frac{\partial U(x,\tau)}{\partial x}\right)d\tau + f(x,t), \quad (2.21)$$

which is (1.14).

2.4 Some features of Models I and II

2.4.1 Existence and uniqueness of the solutions

In the present section we give some features of those types of equations considered in sections 2.1 and 2.2.

As models we shall consider the following two nonlinear partial integro-differential equations:

$$\frac{\partial U}{\partial t} - \frac{\partial}{\partial x}\left[a\left(\int_0^t \left|\frac{\partial U}{\partial x}\right|^2 d\tau\right)\frac{\partial U}{\partial x}\right] = f(x,t) \quad (2.22)$$

and

$$\frac{\partial U}{\partial t} - a\left(\int_0^t \int_0^1 \left|\frac{\partial U}{\partial x}\right|^2 dx d\tau\right) \frac{\partial^2 U}{\partial x^2} = f(x,t). \qquad (2.23)$$

These equations are one-dimensional one-component case of the equations arising at mathematical modeling of process of penetration of an electromagnetic field into a substance with temperature-dependent coefficient of electroconductivity. As we noted in section 2.1, equation (2.22) given here was offered in [187] and [188] at first and then was generalized in numerous works [49], [50], [137], [139], [141], [145], [148], [219], [303], [304], [305], [306], [333], and [334]. Equation (2.23) describes the same process as (2.22) and is first given in [306].

Under certain conditions on the initial data for these equations the existence and uniqueness of the initial-boundary value problems with first kind boundary conditions are given.

For investigations of initial-boundary value problems discussed in the book, we use usual Sobolev spaces $W_p^k(\Omega)$, $\overset{\circ}{W}{}_p^k(\Omega)$, $L_p(\Omega)$, $C^k(\Omega)$, $L_q\left(0,T;\, W_p^k(\Omega)\right)$, $L_q\left(0,T; \overset{\circ}{W}{}_p^k(\Omega)\right)$ and their properties, see, for example, [7], [136], [175], [327], and references therein. Some description of these spaces will be given in section 3.6 as well.

In the domain $Q_T = \Omega \times (0,T)$ of the variables (x,t), where $\Omega = (0,1)$ and T is a fixed positive constant for equations (2.22) and (2.23) we shall consider the following initial-boundary value problem:

$$U(0,t) = U(1,t) = 0, \;\; t \in [0,T], \qquad (2.24)$$

$$U(x,0) = 0, \;\; x \in \overline{\Omega}. \qquad (2.25)$$

We shall search the solution of problem (2.22), (2.24), (2.25) which satisfies the identity

$$\int_{Q_T} \left(\frac{\partial U}{\partial t} + A(U) \right) V dxdt = \int_{Q_T} fV dxdt, \qquad (2.26)$$

where V is any function from the appropriate space (this space for case $a(S) = 1 + S$ will be specified in section 3.6), and

$$A(U) \equiv -\frac{\partial}{\partial x} \left[a \left(\int_0^t \left| \frac{\partial U}{\partial x} \right|^2 d\tau \right) \frac{\partial U}{\partial x} \right].$$

The basic characteristics of equations such as (2.22) is that higher derivatives of the nonlinear term depend on time integral. The coefficients $a = a(S)$ in equation (2.22) contains integral determining nonlocal operator. In the theory of differential equations one often uses the fact that the Volterra's operator of a kind $\int_0^t U d\tau$ is increasing [175]. However, in a combination with derivative on spatial variable this property might be lost. For example, the equation

$$\frac{\partial U}{\partial t} = \frac{\partial}{\partial x} \left[a \left(\int_0^t \frac{\partial U}{\partial x} d\tau \right) \right]$$

is parabolic but if we replace $\int_0^t U d\tau = V$ then the nonlinear equation becomes hyperbolic

$$\frac{\partial^2 V}{\partial t^2} = \frac{\partial}{\partial x} \left[a \left(\frac{\partial V}{\partial x} \right) \right],$$

which generally does not possess global solutions.

The following existence and uniqueness statement takes place for problem (2.22), (2.24), (2.25) (see [138]).

Theorem 2.1 *If*

$$a(S) = (1+S)^p, \ 0 < p \leq 1,$$

and $f \in W_2^1(Q_T)$, *then there exists a unique solution* U *of problem (2.22), (2.24), (2.25) satisfying the following properties:*

$$U \in L_{2p+2}(0,T;\overset{\circ}{W}{}^1_{2p+2}(\Omega)), \ \partial U/\partial t \in L_2(Q_T),$$

$$\frac{\partial^2 U}{\partial x^2} \in L_2(Q_T),$$

$$\sqrt{T-t}\,\frac{\partial^2 U}{\partial t \partial x} \in L_2(Q_T).$$

Theorem 2.1 and some of its generalizations for (2.22), (2.23) type equations and for their multidimensional variants are proved in [137], [138], [139], [141], [145], [187], [188], and [219], with application of Galerkin's and compactness methods [327], [328], and [461].

It should be noted that Theorem 2.1 in case $p = 1$ will be proved completely in section 3.6.

Below we discuss the asymptotic behavior of the solutions of the initial-boundary value problems. The existence result can also be proved based on the a-priori estimates given in the investigation of the asymptotic behavior of the problem for Models I and II.

In [305] and [306] it was shown that equation (2.22) gives rise to coercive monotone operators in the space $\overset{\circ}{W}{}^1_p(0,1;L_2(0,T))$ but not in the traditional $L_2(0,T;\overset{\circ}{W}{}^1_p(0,1))$. In mentioned works by the same scheme the analogous results are obtained for (1.12) type models.

2.4.2 Asymptotic behavior of the solutions as time tends to infinity

This section is devoted to longtime behavior of the solutions of initial-boundary value problems for the nonlinear integro-differential models (2.22) and (2.23). These equations arise at mathematical modeling of process of penetration of an electromagnetic field into a substance, with temperature dependent coefficient of electroconductivity. These equations were derived in sections 2.1 and 2.2.

The initial-boundary value problems with first-type boundary conditions are stated. Investigations, which are made in [217], [218], [228], [231] are given. Let's formulate the statement for asymptotic behavior of the solutions of the problem with homogeneous boundary conditions.

At first for one nonlinear integro-differential problem (2.22) we shall give a-priori estimates of the solutions independent of t. From these estimates the stabilization of the solution follows as $t \to \infty$.

In the cylinder $Q = \Omega \times (0, \infty)$, we consider problem (2.22), (2.24), (2.25).

Let us assume that

$$a(S) \geq a_0 = const > 0.$$

If

$$\beta \leq \frac{2a_0}{C},$$

where C here and below is a positive constant independent of t, the following estimate takes place

$$||U(\cdot, t)|| \leq e^{-\beta t/2} \, ||U_0|| \, .$$

Here $||\cdot||$ is a norm of the space $L_2(\Omega)$, i.e.,

$$||U(\cdot,t)|| = \left(\int_\Omega U^2(x,t)dx\right)^{1/2}.$$

Sometimes, we will use subscript to indicate different kinds of norms.

Thus, for a problem (2.22), (2.24), (2.25) as $t \to \infty$ the stability of the solution has decaying exponential rate in the norm of the space $L_2(\Omega)$. The asymptotic behavior of the solution of problem (2.22), (2.24), (2.25) as $t \to \infty$ can also be proved in the norm of the space $W_2^1(\Omega)$.

Using a technique found in [270], in which the stabilization of the solution of an initial-boundary value problem for the equations of a barotropic viscous liquid is established, the following statement is proved in [217].

Let us assume that $a = a(S)$ satisfies $a(S) \geq a_0 = const > 0$ and the two additional restrictions: $a'(S) \geq 0$ and $a''(S) \leq 0$.

In this case the following a-priori estimates are true for all t [218]:

$$\int_0^t \int_\Omega \left|\frac{\partial U}{\partial x}\right|^2 dx d\tau \leq C,$$

$$\int_0^t \left|\frac{d}{d\tau}\int_\Omega \left|\frac{\partial U}{\partial x}\right|^2 dx\right| d\tau \leq C.$$

Therefore,

$$\int_0^\infty \int_\Omega \left|\frac{\partial U}{\partial x}\right|^2 dx d\tau \leq C,$$

$$\int_0^\infty \left|\frac{d}{d\tau}\int_\Omega \left|\frac{\partial U}{\partial x}\right|^2 dx\right| d\tau \leq C.$$

Now, using well-known fact (see, for example, [25], [270], and [296]) it is clear that

$$||U(\cdot,t)||_{W_2^1(\Omega)} \to 0, \; t \to \infty.$$

Thus, stabilization of the solution of problem (2.22) is proved in the norm of the space $W_2^1(\Omega)$ under the following conditions on coefficient $a = a(S)$:

$$a(S) \geq a_0 = const > 0, \; a'(S) \geq 0, \; a''(S) \leq 0.$$

Note that such type of result is true for problem (2.23)-(2.25) as well.

It is interesting to get the stabilization character in a stronger norm as well. Let us also note that some results about existence, uniqueness, and asymptotic behavior of the solutions for above-discussed integro-differential models and that generalizations with several kinds of initial and boundary conditions are studied in [28], [29], [31], [32], [33], [34], [35], [49], [50], [137], [138], [139], [140], [141], [145], [146], [147], [186], [187], [188], [217], [218], [219], [220], [222], [223], [227], [228], [229], [231], [232], [233], [234], [235], [236], [237], [238], [239], [240], [241], [242], [243], [245], [246], [247], [248], [249], [251], [252], [253], [254], [256], [257], [261], [276], [277], [278], [280], [282], [283], [302], [303], [304], [305], [306], [322], [333], and [334] among others.

2.4.3 Rate of the asymptotic behavior of solutions of Model I

The relations in subsection 2.4.2 give the stabilization of a nonstationary solution with the norm of the space $W_2^1(\Omega)$, but usually do not provide any information on the convergence rate.

Here we rectify this deficiency and give asymptotic behavior to be also confirmed by the numerical experiments in Chapter 4.

Consider the following initial-boundary value problem:

$$\frac{\partial U}{\partial t} = \frac{\partial}{\partial x}\left[a(S)\frac{\partial U}{\partial x}\right], \quad (x,t) \in Q, \tag{2.27}$$

$$U(0,t) = U(1,t) = 0, \quad t \geq 0, \tag{2.28}$$

$$U(x,0) = U_0(x), \quad x \in [0,1], \tag{2.29}$$

where

$$S(x,t) = \int_0^t \left(\frac{\partial U}{\partial x}\right)^2 d\tau. \tag{2.30}$$

Here $a = a(S)$ and $U_0 = U_0(x)$ are given functions of their arguments.

Let us estimate the rate of asymptotic behavior of the solution. In particular, let us show that the convergence occurring in the relations in subsection 2.4.2 is also exponential.

Note that, in this section, we derive asymptotics with indication of order for $\|\partial U/\partial t\|$ as well.

Let us note once again that below in this and in the next sections C denote positive constants independent of t.

Theorem 2.2 *If $a(S) \geq a_0 = const > 0$, $a'(S) \geq 0$, $a''(S) \leq 0$, and $U_0 \in W_2^2(0,1) \cap \overset{\circ}{W}{}_2^k(0,1)$, then the solution of problem (2.27)-(2.30) satisfies the relation*

$$\left\|\frac{\partial U}{\partial x}\right\| + \left\|\frac{\partial U}{\partial t}\right\| \leq C \exp\left(-\frac{a_0}{2}t\right).$$

Proof. We multiply (2.27) by $\partial U/\partial t$ and integrate the resulting relation over the interval $[0,1]$. By using the formula of integration by parts and taking into account the boundary conditions

(2.28), we obtain

$$\left\|\frac{\partial U}{\partial t}\right\|^2 + \frac{1}{2}\int_0^1 a(S)\,\frac{\partial}{\partial t}\left(\frac{\partial U}{\partial x}\right)^2 dx = 0. \tag{2.31}$$

Let us differentiate relation (2.27) with respect to t:

$$\frac{\partial^2 U}{\partial t^2} - \frac{\partial}{\partial x}\left[\frac{\partial a(S)}{\partial t}\,\frac{\partial U}{\partial x} + a(S)\frac{\partial^2 U}{\partial t\,\partial x}\right] = 0. \tag{2.32}$$

We multiply (2.32) by U and integrate the resulting relation over the interval $[0, 1]$:

$$\begin{aligned}&\int_0^1 \frac{\partial^2 U}{\partial t^2}\,U\,dx + \int_0^1 \frac{\partial a(S)}{\partial t}\left(\frac{\partial U}{\partial x}\right)^2 dx\\ &+\frac{1}{2}\int_0^1 a(S)\frac{\partial}{\partial t}\left(\frac{\partial U}{\partial x}\right)^2 dx = 0.\end{aligned} \tag{2.33}$$

It follows from (2.31) and (2.33) that

$$\begin{aligned}&\int_0^1 \frac{\partial^2 U}{\partial t^2}\,U\,dx + \left\|\frac{\partial U}{\partial t}\right\|^2 + \int_0^1 a(S)\,\frac{\partial}{\partial t}\left(\frac{\partial U}{\partial x}\right)^2 dx\\ &+\int_0^1 \frac{\partial a(S)}{\partial t}\left(\frac{\partial U}{\partial x}\right)^2 dx = 0,\end{aligned}$$

which, in view of the relation

$$\int_0^1 \frac{\partial^2 U}{\partial t^2}\,U\,dx + \left\|\frac{\partial U}{\partial t}\right\|^2 = \frac{1}{2}\,\frac{d^2}{dt^2}\,\|U\|^2,$$

acquires the form

$$\frac{1}{2}\,\frac{d^2}{dt^2}\|U\|^2 + \frac{d}{dt}\int_0^1 a(S)\left(\frac{\partial U}{\partial x}\right)^2 dx = 0. \tag{2.34}$$

Let us multiply (2.32) by $\partial U/\partial t$ and integrate the resulting relation over the interval $[0,1]$. By using the formula of integration by parts, the boundary conditions (2.28), and Poincaré's inequality

$$\left\|\frac{\partial U}{\partial t}\right\| \leq \left\|\frac{\partial^2 U}{\partial t\,\partial x}\right\|,$$

we obtain

$$\frac{d}{dt}\left\|\frac{\partial U}{\partial t}\right\|^2 + 2a_0\left\|\frac{\partial U}{\partial t}\right\|^2 + \frac{1}{2}\int_0^1 a'(S)\,\frac{\partial}{\partial t}\left(\frac{\partial U}{\partial x}\right)^4 dx \leq 0. \quad (2.35)$$

Multiplying (2.27) by U and integrating the resulting relation over the interval $[0,1]$, we obtain

$$\frac{1}{2}\frac{d}{dt}\,\|U\|^2 + \int_0^1 a(S)\left(\frac{\partial U}{\partial x}\right)^2 dx = 0, \quad (2.36)$$

which, together with Poincaré's inequality and the condition $a(S) \geq a_0$, implies that

$$\frac{1}{2}\frac{d}{dt}\|U\|^2 + a_0\|U\|^2 \leq 0. \quad (2.37)$$

We multiply relations (2.34), (2.36), and (2.37) by γ, α, and β, respectively, where α, β, and γ are positive constants. By summing the resulting relations and inequality (2.35), we obtain

$$\frac{\alpha+\beta}{2}\,\frac{d}{dt}\|U\|^2 + \alpha\int_0^1 a(S)\left(\frac{\partial U}{\partial x}\right)^2 dx + \beta a_0\|U\|^2$$

$$+\frac{\gamma}{2}\,\frac{d^2}{dt^2}\,\|U\|^2 + \gamma\,\frac{d}{dt}\int_0^1 a(S)\left(\frac{\partial U}{\partial x}\right)^2 dx + \frac{d}{dt}\left\|\frac{\partial U}{\partial t}\right\|^2$$

$$+2a_0\left\|\frac{\partial U}{\partial t}\right\|^2 + \frac{1}{2}\int_0^1 a'(S)\,\frac{\partial}{\partial t}\left(\frac{\partial U}{\partial x}\right)^4 dx \leq 0,$$

which, after simple manipulations, acquires the form

$$\begin{aligned}
&\frac{\alpha+\beta}{2}\left[\frac{d}{dt}\|U\|^2+\frac{2\beta a_0}{\alpha+\beta}\|U\|^2\right]\\
&+\gamma\left[\frac{d}{dt}\int_0^1 a(S)\left(\frac{\partial U}{\partial x}\right)^2 dx+\frac{\alpha}{\gamma}\int_0^1 a(S)\left(\frac{\partial U}{\partial x}\right)^2 dx\right]\\
&+\frac{d}{dt}\left\|\frac{\partial U}{\partial t}\right\|^2+2a_0\left\|\frac{\partial U}{\partial t}\right\|^2\\
&+\frac{\gamma}{2}\frac{d^2}{dt^2}\|U\|^2+\frac{1}{2}\int_0^1 a'(S)\frac{\partial}{\partial t}\left(\frac{\partial U}{\partial x}\right)^4 dx\le 0.
\end{aligned}\tag{2.38}$$

Suppose that

$$2\beta a_0/(\alpha+\beta)=\alpha/\gamma=a_0,$$

then

$$\alpha=\beta=(\alpha+\beta)/2=a_0\gamma$$

and inequality (2.38) can be represented in the form

$$\begin{aligned}
&a_0\gamma\,\frac{d}{dt}[\exp(a_0t)\|U\|^2]+\gamma\,\frac{d}{dt}\left[\exp(a_0t)\int_0^1 a(S)\left(\frac{\partial U}{\partial x}\right)^2 dx\right]\\
&+\frac{d}{dt}\left[\exp(a_0t)\left\|\frac{\partial U}{\partial t}\right\|^2\right]+\exp(a_0t)\left[a_0\left\|\frac{\partial U}{\partial t}\right\|^2\right.\\
&\left.+\frac{\gamma}{2}\frac{d^2}{dt^2}\|U\|^2+\frac{1}{2}\int_0^1 a'(S)\,\frac{\partial}{\partial t}\left(\frac{\partial U}{\partial x}\right)^4 dx\right]\le 0.
\end{aligned}$$

By integrating the resulting inequality over the closed interval $[0,\ t]$ and taking into account the initial condition (2.29), we obtain the inequality

$$
\begin{aligned}
&a_0\gamma \exp(a_0 t)\|U\|^2 + \gamma \exp(a_0 t)\int_0^1 a(S)\left(\frac{\partial U}{\partial x}\right)^2 dx \\
&+\exp(a_0 t)\left\|\frac{\partial U}{\partial t}\right\|^2 + a_0\int_0^t \exp(a_0\tau)\left\|\frac{\partial U}{\partial \tau}\right\|^2 d\tau \\
&+\frac{\gamma}{2}\int_0^t \exp(a_0\tau)\frac{d^2}{d\tau^2}\|U\|^2 d\tau \\
&+\frac{1}{2}\int_0^t \exp(a_0\tau)\int_0^1 a'(S)\frac{\partial}{\partial \tau}\left(\frac{\partial U}{\partial x}\right)^4 dx\, d\tau \le C.
\end{aligned}
\tag{2.39}
$$

Note that

$$
\begin{aligned}
&\int_0^t \exp(a_0\tau)\int_0^1 a'(S)\frac{\partial}{\partial \tau}\left(\frac{\partial U}{\partial x}\right)^4 dx\, d\tau \\
&= \exp(a_0 t)\int_0^1 a'(S)\left(\frac{\partial U}{\partial x}\right)^4 dx \\
&- \int_0^1 a'(S)\left(\frac{\partial U}{\partial x}\right)^4 dx\bigg|_{t=0} \\
&-a_0\int_0^t \exp(a_0\tau)\int_0^1 a'(S)\left(\frac{\partial U}{\partial x}\right)^4 dx\, d\tau \\
&- \int_0^t \exp(a_0\tau)\int_0^1 a''(S)\left(\frac{\partial U}{\partial x}\right)^6 dx\, d\tau \\
&\ge -a_0\int_0^t \exp(a_0\tau)\int_0^1 a'(S)\left(\frac{\partial U}{\partial x}\right)^4 dx\, d\tau - C,
\end{aligned}
\tag{2.40}
$$

and we have the identity

$$\int_0^t \exp(a_0\tau)\frac{d^2}{d\tau^2}\|U\|^2 d\tau = \exp(a_0 t)\frac{d}{dt}\|U\|^2$$

$$-a_0\int_0^t \exp(a_0\tau)\frac{d}{d\tau}\|U\|^2 d\tau - \frac{d}{dt}\|U\|^2\bigg|_{t=0}$$

$$= 2\exp(a_0 t)\int_0^1 U\frac{\partial U}{\partial t}\,dx - a_0\exp(a_0 t)\|U\|^2$$

$$+a_0^2\int_0^t \exp(a_0\tau)\|U\|^2 d\tau + a_0\|U_0\|^2 - \frac{d}{dt}\|U\|^2\bigg|_{t=0}.$$

By using the ε-inequality in the last relation, we obtain

$$\int_0^t \exp(a_0\tau)\frac{d^2}{d\tau^2}\|U\|^2 d\tau$$

$$\geq -(\varepsilon + a_0)\exp(a_0 t)\|U\|^2 \tag{2.41}$$

$$-\frac{1}{\varepsilon}\exp(a_0 t)\left\|\frac{\partial U}{\partial t}\right\|^2 - C.$$

We multiply relation (2.33) by $\mu\exp(a_0 t)$, where μ is a positive constant, and integrate the resulting relation over the interval $[0,t]$:

$$\mu\int_0^t \exp(a_0\tau)\int_0^1 \frac{\partial^2 U}{\partial\tau^2}U\,dx\,d\tau$$

$$+\mu\int_0^t \exp(a_0\tau)\int_0^1 a'(S)\left(\frac{\partial U}{\partial x}\right)^4 dx\,d\tau$$

$$+\frac{\mu}{2}\int_0^t \exp(a_0\tau)\int_0^1 a(S)\frac{\partial}{\partial t}\left(\frac{\partial U}{\partial x}\right)^2 dx\,d\tau = 0.$$

Hence, by performing integration by parts, we have

$$\mu \exp(a_0 t) \int_0^1 \frac{\partial U}{\partial t} U \, dx$$

$$-a_0 \mu \int_0^t \exp(a_0 \tau) \int_0^1 \frac{\partial U}{\partial \tau} U \, dx \, d\tau$$

$$-\mu \int_0^t \exp(a_0 \tau) \left\| \frac{\partial U}{\partial \tau} \right\|^2 d\tau$$

$$+\mu \int_0^t \exp(a_0 \tau) \int_0^1 a'(S) \left(\frac{\partial U}{\partial x} \right)^4 dx \, d\tau$$

$$+\frac{\mu}{2} \int_0^t \exp(a_0 \tau) \int_0^1 a(S) \frac{\partial}{\partial t} \left(\frac{\partial U}{\partial x} \right)^2 dx \, d\tau = C.$$

By using again the ε-inequality with parameter ε' and taking into account (2.31) and (2.36), we obtain

$$-\varepsilon' \mu \exp(a_0 t) \|U\|^2 - \frac{\mu \exp(a_0 t)}{4\varepsilon'} \left\| \frac{\partial U}{\partial t} \right\|^2$$

$$+a_0 \mu \int_0^t \exp(a_0 \tau) \int_0^1 a(S) \left(\frac{\partial U}{\partial x} \right)^2 dx \, d\tau$$

$$-\mu \int_0^t \exp(a_0 \tau) \left\| \frac{\partial U}{\partial \tau} \right\|^2 d\tau \qquad (2.42)$$

$$+\mu \int_0^t \exp(a_0 \tau) \int_0^1 a'(S) \left(\frac{\partial U}{\partial x} \right)^4 dx \, d\tau$$

$$-\mu \int_0^t \exp(a_0 \tau) \left\| \frac{\partial U}{\partial \tau} \right\|^2 d\tau \le C.$$

From (2.37) with $\eta > 0$ we obtain the inequality

$$\eta \exp(a_0 t)\|U\|^2 \leq C. \tag{2.43}$$

By combining inequalities (2.39)-(2.43), we obtain

$$\exp(a_0 t)\left[\left(a_0\gamma - \frac{\gamma(\varepsilon + a_0)}{2} - \mu\varepsilon' + \eta\right)\|U\|^2\right.$$

$$\left. +\gamma a_0\left\|\frac{\partial U}{\partial x}\right\|^2 + \left(1 - \frac{\gamma}{2\varepsilon} - \frac{\mu}{4\varepsilon'}\right)\left\|\frac{\partial U}{\partial t}\right\|^2\right]$$

$$+(a_0 - 2\mu)\int_0^t \exp(a_0\tau)\left\|\frac{\partial U}{\partial \tau}\right\|^2 d\tau$$

$$+\left(\mu - \frac{a_0}{2}\right)\int_0^t \exp(a_0\tau)\int_0^1 a'(S)\left(\frac{\partial U}{\partial x}\right)^4 dx\, d\tau$$

$$+a_0^2\mu\int_0^t \exp(a_0\tau)\left\|\frac{\partial U}{\partial x}\right\|^2 d\tau \leq C.$$

To complete the proof of Theorem 2.2, it remains to show that the coefficients occurring on the left-hand side in the last inequality can be chosen in an appropriate way. For example, this can be achieved by an appropriate choice of the free parameters as follows: $\mu = a_0/2$, $\varepsilon = \varepsilon' = a_0$, $\gamma < 7a_0/4$, and $\eta \geq a_0^2/2$.

The proof of Theorem 2.2 is complete.

Let us note that asymptotic behavior of the solution of problem (2.27)-(2.30) is studied in the norm of the space $C^1(0,1)$ too (see [240]). For averaged model such type result is given in the next section.

2.4.4 Rate of the asymptotic behavior of solutions of Model II

Consider the following initial-boundary value problem:

$$\frac{\partial U}{\partial t} = a(S)\frac{\partial^2 U}{\partial x^2}, \quad (x,t) \in Q, \tag{2.44}$$

$$U(0,t) = U(1,t) = 0, \quad t \geq 0, \tag{2.45}$$

$$U(x,0) = U_0(x), \quad x \in [0,1], \tag{2.46}$$

where

$$S(t) = \int_0^t \int_0^1 \left(\frac{\partial U}{\partial x}\right)^2 dx d\tau, \tag{2.47}$$

$a(S) = (1+S)^p$, $p > 0$; $U_0 = U_0(x)$ is a given function.

The main purpose of this section is to prove the following statement.

Theorem 2.3 *If* $a(S) = (1+S)^p$, $p > 0$; $U_0 \in W_2^4(0,1) \cap \overset{\circ}{W}{}_2^1(0,1)$, *then for the solution of problem (2.44)-(2.47) the following estimates hold:*

$$\left|\frac{\partial U(x,t)}{\partial x}\right| \leq C \exp\left(-\frac{t}{2}\right), \quad \left|\frac{\partial U(x,t)}{\partial t}\right| \leq C \exp\left(-\frac{t}{2}\right).$$

To prove Theorem 2.3 we need some auxiliary statements.

Theorem 2.4 *For the solution of problem (2.44)-(2.47) the following estimate is true*

$$\|U\| + \left\|\frac{\partial U}{\partial x}\right\| \leq C \exp\left(-\frac{t}{2}\right).$$

Proof. Let us multiply equation (2.44) by U and integrate over $[0,1]$. After integrating by parts and using the boundary conditions (2.45), we get

$$\frac{1}{2}\frac{d}{dt}\|U\|^2 + \int_0^1 (1+S)^p \left(\frac{\partial U}{\partial x}\right)^2 dx = 0.$$

From this, taking into account the relation $(1+S)^p \geq 1$ and Poincaré's inequality

$$\|U\| \leq \left\|\frac{\partial U}{\partial x}\right\|,$$

we obtain:

$$\begin{aligned} &\frac{1}{2}\frac{d}{dt}\|U\|^2 + \left\|\frac{\partial U}{\partial x}\right\|^2 \leq 0, \\ &\frac{1}{2}\frac{d}{dt}\|U\|^2 + \|U\|^2 \leq 0. \end{aligned} \tag{2.48}$$

Let us multiply equation (2.44) by $\partial^2 U/\partial x^2$ and integrate over [0, 1]. Using again formula of integrating by parts and boundary conditions (2.45), we get:

$$\left.\frac{\partial U}{\partial t}\frac{\partial U}{\partial x}\right|_0^1 - \int_0^1 \frac{\partial^2 U}{\partial x \partial t}\frac{\partial U}{\partial x} dx = \int_0^1 (1+S)^p \left(\frac{\partial^2 U}{\partial x^2}\right)^2 dx,$$

$$\frac{1}{2}\frac{d}{dt}\left\|\frac{\partial U}{\partial x}\right\|^2 + (1+S)^p \left\|\frac{\partial^2 U}{\partial x^2}\right\|^2 = 0, \tag{2.49}$$

or

$$\frac{d}{dt}\left\|\frac{\partial U}{\partial x}\right\|^2 \leq 0. \tag{2.50}$$

Combining (2.48) and (2.50), we deduce

$$\frac{d}{dt}\|U\|^2 + \|U\|^2 + \frac{d}{dt}\left\|\frac{\partial U}{\partial x}\right\|^2 + \left\|\frac{\partial U}{\partial x}\right\|^2 \leq 0.$$

After multiplying by the function $\exp(t)$ last inequality gives

$$\frac{d}{dt}\left(\exp(t)\,\|U\|^2\right) + \frac{d}{dt}\left(\exp(t)\left\|\frac{\partial U}{\partial x}\right\|^2\right) \le 0$$

or

$$\frac{d}{dt}\left[\exp(t)\left(\|U\|^2 + \left\|\frac{\partial U}{\partial x}\right\|^2\right)\right] \le 0.$$

This inequality immediately proves Theorem 2.4.

Note that Theorem 2.4 gives exponential stabilization of the solution of problem (2.44)-(2.47) in the norm of the space $W_2^1(0,1)$. Note also that Theorem 2.4 is true for any $a = a(S) \ge a_0 = const > 0$.

As we already said we want to prove Theorem 2.3, i.e., to prove the asymptotic behavior with the stabilization rate in $C^1(0,1)$. To this end we need to prove the following statement.

Theorem 2.5 *For the solution of problem (2.44)-(2.47) the following estimate is true*

$$\left\|\frac{\partial U(x,t)}{\partial t}\right\| \le C\exp\left(-\frac{t}{2}\right).$$

Proof. Let us differentiate equation (2.44) with respect to t,

$$\frac{\partial^2 U}{\partial t^2} = (1+S)^p \frac{\partial^3 U}{\partial x^2 \partial t} + p(1+S)^{p-1}\int_0^1 \left(\frac{\partial U}{\partial x}\right)^2 dx \frac{\partial^2 U}{\partial x^2}, \quad (2.51)$$

and multiply (2.51) by $\partial U/\partial t$ and integrate over [0, 1]. We deduce

$$\frac{1}{2}\frac{d}{dt}\int_0^1 \left(\frac{\partial U}{\partial t}\right)^2 dx + (1+S)^p \int_0^1 \left(\frac{\partial^2 U}{\partial x \partial t}\right)^2 dx$$

$$+p(1+S)^{p-1}\int_0^1 \left(\frac{\partial U}{\partial x}\right)^2 dx \int_0^1 \frac{\partial U}{\partial x}\frac{\partial^2 U}{\partial x \partial t} dx = 0$$

or

$$\frac{d}{dt}\int_0^1 \left(\frac{\partial U}{\partial t}\right)^2 dx + 2(1+S)^p \int_0^1 \left(\frac{\partial^2 U}{\partial x \partial t}\right)^2 dx$$

$$= -2p(1+S)^{p-1} \int_0^1 \left(\frac{\partial U}{\partial x}\right)^2 dx \int_0^1 \frac{\partial U}{\partial x}\frac{\partial^2 U}{\partial x \partial t} dx.$$

Let us estimate the right-hand side of the last equality

$$-2p(1+S)^{p-1} \int_0^1 \left(\frac{\partial U}{\partial x}\right)^2 dx \int_0^1 \frac{\partial U}{\partial x}\frac{\partial^2 U}{\partial x \partial t} dx$$

$$= -2\int_0^1 \left\{ p(1+S)^{p/2-1} \left[\int_0^1 \left(\frac{\partial U}{\partial x}\right)^2 dx\right] \frac{\partial U}{\partial x} \right\}$$

$$\times \left\{ (1+S)^{p/2} \frac{\partial^2 U}{\partial x \partial t} \right\} dx.$$

From this, using the Schwarz's inequality, we get

$$\frac{d}{dt}\int_0^1 \left(\frac{\partial U}{\partial t}\right)^2 dx + 2(1+S)^p \int_0^1 \left(\frac{\partial^2 U}{\partial x \partial t}\right)^2 dx$$

$$\leq (1+S)^p \int_0^1 \left(\frac{\partial^2 U}{\partial x \partial t}\right)^2 dx$$

$$+p^2(1+S)^{p-2} \left[\int_0^1 \left(\frac{\partial U}{\partial x}\right)^2 dx\right]^2 \int_0^1 \left(\frac{\partial U}{\partial x}\right)^2 dx$$

or

$$\begin{aligned}\frac{d}{dt}\int_0^1 \left(\frac{\partial U}{\partial t}\right)^2 dx + (1+S)^p \int_0^1 \left(\frac{\partial^2 U}{\partial x \partial t}\right)^2 dx \\ \leq p^2(1+S)^{p-2}\left[\int_0^1 \left(\frac{\partial U}{\partial x}\right)^2 dx\right]^3.\end{aligned} \tag{2.52}$$

Note that Theorem 2.4 enables us to estimate function S

$$S(t) = \int_0^t \int_0^1 \left(\frac{\partial U}{\partial x}\right)^2 dx d\tau = \int_0^t \left\|\frac{\partial U}{\partial x}\right\|^2 d\tau$$

$$\leq C\int_0^t \exp(-\tau)d\tau \leq C.$$

So, we have

$$1 \leq 1 + S(t) \leq C. \tag{2.53}$$

Using Poincaré's inequality

$$\left\|\frac{\partial U}{\partial t}\right\| \leq \left\|\frac{\partial^2 U}{\partial t\, \partial x}\right\|,$$

Theorem 2.4, and relation (2.53), from (2.52) we get

$$\frac{d}{dt}\int_0^1 \left(\frac{\partial U}{\partial t}\right)^2 dx + \int_0^1 \left(\frac{\partial U}{\partial t}\right)^2 dx \leq C\exp(-3\tau).$$

Multiplying last inequality by the function $\exp(t)$, we arrive at

$$\frac{d}{dt}\left(\exp(t)\left\|\frac{\partial U}{\partial t}\right\|^2\right) \leq C\exp(-2t).$$

From this we deduce

$$\exp(t)\left\|\frac{\partial U}{\partial t}\right\|^2 \leq C\int_0^t \exp(-2\tau)d\tau.$$

Therefore, Theorem 2.5 is proved.

Now we prove the main theorem of this section.

Proof of Theorem 2.3. Let us estimate $\partial^2 U/\partial x^2$ in the norm of the space $L_1(0,1)$. From (2.44) we have

$$\frac{\partial^2 U}{\partial x^2} = (1+S)^{-p}\frac{\partial U}{\partial t}. \tag{2.54}$$

So, applying again Schwarz's inequality, Theorem 2.5, and estimate (2.54), we derive

$$\int_0^1 \left|\frac{\partial^2 U}{\partial x^2}\right| dx \le \left[\int_0^1 (1+S)^{-2p} dx\right]^{1/2} \left[\int_0^1 \left|\frac{\partial U}{\partial t}\right|^2 dx\right]^{1/2}$$

$$\le C\exp\left(-\frac{t}{2}\right).$$

From this, taking into account the relation

$$\frac{\partial U(x,t)}{\partial x} = \int_0^1 \frac{\partial U(y,t)}{\partial y} dy + \int_0^1 \int_y^x \frac{\partial^2 U(\xi,t)}{\partial \xi^2} d\xi dy$$

and the boundary conditions (2.45), it follows that

$$\left|\frac{\partial U(x,t)}{\partial x}\right| = \left|\int_0^1 \int_y^x \frac{\partial^2 U(\xi,t)}{\partial \xi^2} d\xi dy\right|$$

$$\le \int_0^1 \left|\frac{\partial^2 U(y,t)}{\partial y^2}\right| dy \le C\exp\left(-\frac{t}{2}\right).$$

Now let us estimate $\partial U/\partial t$ in the norm of the space $C(0,1)$. Let us multiply equation (2.44) by $\partial^3 U/\partial x^2 \partial t$ and integrate over [0, 1]. After using the formula of integrating by parts and boundary conditions (2.45), we get

$$\frac{\partial U}{\partial t}\frac{\partial^2 U}{\partial x \partial t}\Big|_0^1 - \left\|\frac{\partial^2 U}{\partial x \partial t}\right\|^2 = (1+S)^p \int_0^1 \frac{\partial^2 U}{\partial x^2}\frac{\partial^3 U}{\partial x^2 \partial t} dx, \tag{2.55}$$

$$\frac{1}{2}(1+S)^p \frac{d}{dt} \left\| \frac{\partial^2 U}{\partial x^2} \right\|^2 + \left\| \frac{\partial^2 U}{\partial x \partial t} \right\|^2 = 0$$

or

$$\frac{d}{dt} \left\| \frac{\partial^2 U}{\partial x^2} \right\|^2 \leq 0. \tag{2.56}$$

Note that from (2.55) we have

$$\begin{aligned} \left\| \frac{\partial^2 U}{\partial x \partial t} \right\|^2 &\leq (1+S)^p \left[\int_0^1 \left(\frac{\partial^2 U}{\partial x^2} \right)^2 dx \right]^{1/2} \\ &\quad \times \left[\int_0^1 \left(\frac{\partial^3 U}{\partial x^2 \partial t} \right)^2 dx \right]^{1/2} \qquad (2.57) \\ &\leq \frac{1}{2}(1+S)^p \left\| \frac{\partial^2 U}{\partial x^2} \right\|^2 + \frac{1}{2}(1+S)^p \left\| \frac{\partial^3 U}{\partial x^2 \partial t} \right\|^2 . \end{aligned}$$

Let us multiply (2.51) by $\partial^3 U/\partial x^2 \partial t$ and integrate over $[0,\ 1]$

$$\begin{aligned} &\left. \frac{\partial^2 U}{\partial t^2} \frac{\partial^2 U}{\partial x \partial t} \right|_0^1 - \int_0^1 \frac{\partial^3 U}{\partial x \partial t^2} \frac{\partial^2 U}{\partial x \partial t} dx \\ &= (1+S)^p \left\| \frac{\partial^3 U}{\partial x^2 \partial t} \right\|^2 \\ &+ p(1+S)^{p-1} \int_0^1 \left(\frac{\partial U}{\partial x} \right)^2 dx \int_0^1 \frac{\partial^2 U}{\partial x^2} \frac{\partial^3 U}{\partial x^2 \partial t} dx. \end{aligned}$$

Taking into account boundary conditions (2.45), we obtain

$$\frac{d}{dt}\left\|\frac{\partial^2 U}{\partial x\partial t}\right\|^2 + 2(1+S)^p\left\|\frac{\partial^3 U}{\partial x^2\partial t}\right\|^2$$

$$= -2\int_0^1 \left\{p(1+S)^{p/2-1}\left[\int_0^1\left(\frac{\partial U}{\partial x}\right)^2 dx\right]\frac{\partial^2 U}{\partial x^2}\right\}$$

$$\times\left\{(1+S)^{p/2}\frac{\partial^3 U}{\partial x^2\partial t}\right\}dx.$$

From this, using Schwarz's inequality once more, we receive

$$\frac{d}{dt}\left\|\frac{\partial^2 U}{\partial x\partial t}\right\|^2 + 2(1+S)^p\left\|\frac{\partial^3 U}{\partial x^2\partial t}\right\|^2$$

$$\le (1+S)^p\left\|\frac{\partial^3 U}{\partial x^2\partial t}\right\|^2$$

$$+p^2(1+S)^{p-2}\left[\int_0^1\left(\frac{\partial U}{\partial x}\right)^2 dx\right]^2\int_0^1\left(\frac{\partial^2 U}{\partial x^2}\right)^2 dx$$

or

$$\frac{d}{dt}\left\|\frac{\partial^2 U}{\partial x\partial t}\right\|^2 + (1+S)^p\left\|\frac{\partial^3 U}{\partial x^2\partial t}\right\|^2$$

$$\le p^2(1+S)^{p-2}\left[\int_0^1\left(\frac{\partial U}{\partial x}\right)^2 dx\right]^2\int_0^1\left(\frac{\partial^2 U}{\partial x^2}\right)^2 dx.$$

Using Theorem 2.4, relations (2.53), (2.54), and Theorem 2.5, we arrive at

$$\frac{d}{dt}\left\|\frac{\partial^2 U}{\partial x\partial t}\right\|^2 + (1+S)^p\left\|\frac{\partial^3 U}{\partial x^2\partial t}\right\|^2 \le C\exp(-3t). \qquad (2.58)$$

Combining (2.48), (2.49), and (2.56)-(2.58), we get

$$\|U\|^2 + \frac{d}{dt}\|U\|^2 + \left\|\frac{\partial U}{\partial x}\right\|^2 + \frac{d}{dt}\left\|\frac{\partial U}{\partial x}\right\|^2$$

$$+2(1+S)^p\left\|\frac{\partial^2 U}{\partial x^2}\right\|^2 + \frac{d}{dt}\left\|\frac{\partial^2 U}{\partial x^2}\right\|^2$$

$$+\left\|\frac{\partial^2 U}{\partial x\partial t}\right\|^2 + \frac{d}{dt}\left\|\frac{\partial^2 U}{\partial x\partial t}\right\|^2 + (1+S)^p\left\|\frac{\partial^3 U}{\partial x^2\partial t}\right\|^2$$

$$\leq \frac{1}{2}(1+S)^p\left\|\frac{\partial^2 U}{\partial x^2}\right\|^2$$

$$+\frac{1}{2}(1+S)^p\left\|\frac{\partial^3 U}{\partial x^2\partial t}\right\|^2 + C\exp(-3t).$$

After simple transformation, keeping in mind estimate (2.53), we deduce

$$\|U\|^2 + \frac{d}{dt}\|U\|^2 + \left\|\frac{\partial U}{\partial x}\right\|^2 + \frac{d}{dt}\left\|\frac{\partial U}{\partial x}\right\|^2$$

$$+\left\|\frac{\partial^2 U}{\partial x^2}\right\|^2 + \frac{d}{dt}\left\|\frac{\partial^2 U}{\partial x^2}\right\|^2$$

$$+\left\|\frac{\partial^2 U}{\partial x\partial t}\right\|^2 + \frac{d}{dt}\left\|\frac{\partial^2 U}{\partial x\partial t}\right\|^2 \leq C\exp(-3t).$$

From this, after multiplying by the function $\exp(t)$, we get

$$\frac{d}{dt}\left[\exp(t)\left(\|U\|^2 + \left\|\frac{\partial U}{\partial x}\right\|^2 + \left\|\frac{\partial^2 U}{\partial x^2}\right\|^2 + \left\|\frac{\partial^2 U}{\partial x\partial t}\right\|^2\right)\right]$$

$$\leq C\exp(-2t)$$

or

$$\|U\|^2 + \left\|\frac{\partial U}{\partial x}\right\|^2 + \left\|\frac{\partial^2 U}{\partial x^2}\right\|^2 + \left\|\frac{\partial^2 U}{\partial x \partial t}\right\|^2 \leq C\exp(-t).$$

At last, taking into account the relation

$$\frac{\partial U(x,t)}{\partial t} = \int_0^1 \frac{\partial U(y,t)}{\partial t} dy + \int_0^1 \int_y^x \frac{\partial^2 U(\xi,t)}{\partial \xi \partial t} d\xi dy$$

and Theorem 2.5, we obtain

$$\left|\frac{\partial U(x,t)}{\partial t}\right| \leq \left[\int_0^1 \left(\frac{\partial U(x,t)}{\partial t}\right)^2 dx\right]^{1/2} + \left[\int_0^1 \left(\frac{\partial^2 U(x,t)}{\partial x \partial t}\right)^2 dx\right]^{1/2}$$

$$\leq C\exp\left(-\frac{t}{2}\right).$$

So, the main Theorem 2.3 is proved.

Now, the existence of globally defined solutions of problem (2.44)-(2.47) can be re-obtained by a routine procedure. One first establishes the existence of local solutions on a maximal time interval and then deduces from the obtained estimates that this solution cannot escape in a finite time (see, for example, [133], [327], [461]).

One must also note that one approach of proof of theorem of existence, based on Galerkin's and compactness methods for first boundary problem for equation (2.27) for a case $a(S) = 1 + S$ is given in section 3.6.

2.5 Some features of Model III

In this section some qualitative characteristics of model (2.21) are considered.

The model (2.21) behaves the same way as the damped nonlinear wave equation

$$\frac{\partial^2 U(x,t)}{\partial t^2} + \alpha \frac{\partial U(x,t)}{\partial t} - \frac{\partial}{\partial x}\sigma\left(\frac{\partial U(x,t)}{\partial x}\right) = g(x,t),$$

with the following initial conditions:

$$U(x,0) = U_0(x),$$

$$\frac{\partial U(x,0)}{\partial t} = U_1(x),$$

where α is some properly chosen positive constant depending on the kernel a. This was shown by MacCamy [345] and [346].

2.5.1 Existence and uniqueness

MacCamy [345] proved the existence of a global classical solution for (2.21) subject to the initial and boundary conditions:

$$U(0,t) = U(1,t) = 0, \quad t \geq 0, \tag{2.59}$$

$$U(x,0) = U_0(x), \quad x \in [0,\ 1], \tag{2.60}$$

for sufficiently small data. To be more precise, we quote the existence and uniqueness statement and define smallness of the data.

Theorem 2.6 *Suppose that for some $N \geq 4$ the kernel a, the functions f, and $U_0(x)$ satisfy:*

$$t^j a^{(k)} \in L_1(0,\infty),\ k = 0,1,2,\ j \leq 3 + N, \tag{2.61}$$

$$f(0,t) \equiv f(1,t) \equiv \frac{\partial^2 f(0,t)}{\partial x^2} \equiv \frac{\partial^2 f(1,t)}{\partial x^2},$$

$$t^j\bar{f}(t) \in L_1(0,\infty) \cap L_2(0,\infty) \cap L_\infty(0,\infty),\ j \leq N,$$

where

$$\bar{f}(t) = \sup_{x\in[0,1]} (|f(x,t)|, |f_t(x,t)|),$$

$$U_0(x) \in C^3[0,1],\ \ U_0(0) = U_0(1) = U''(0) = U''(1) = 0.$$

Then if D is sufficiently small there exists a unique solution of the initial-boundary value problem (2.21), (2.59), and (2.60), where D is given by

$$D = \sup_{x\in[0,1]} \sum_{j=0}^{2} \left| \frac{d^jU_0(x)}{dx^j} \right| + \sum_{j=0}^{N} ||t^j\bar{f}(t)||_{L_1(0,\infty)}$$

$$+ \sum_{j=0}^{N} \left(||t^j\bar{f}(t)||_{L_2(0,\infty)} + ||t^j\bar{f}(t)||_{L_\infty(0,\infty)} \right).$$

Note that in the case of nonlinear viscoelasticity, one requires the kernel a to satisfy $a = a_\infty + b$, with $a_\infty > 0$ and $b \in L_1(0,\infty)$. The theory developed by MacCamy [345] does not apply to this case.

MacCamy followed Nishida [382] and used Riemann invariants. Therefore, the results are restricted to one space dimension. Dafermos and Nohel [122] has used energy estimates similar to those employed by Matsumura [352] to generalize the results obtained by MacCamy.

Staffans [438] has combined the ideas of energy method but instead of time derivatives he uses differentiation with respect to space variable x. Thus his assumptions are slightly weaker.

2.5.2 Asymptotic behavior

Under the additional conditions on the kernel a

$$a \in C^2[0,\infty), \quad a(0) > 0, \quad \dot{a}(0) < 0,$$

and the conditions on the nonlinear function σ:

$$\sigma \in C^2(-\infty,\infty), \qquad \sigma(0) = 0, \tag{2.62}$$

$$0 \leq \mu \leq \sigma'(\xi) \leq \gamma, \qquad \forall \xi, \tag{2.63}$$

MacCamy [345] has proved the following result.

Theorem 2.7 *If U is a solution to the initial-boundary value problem (2.21), (2.59) and (2.60), then*

$$\lim_{t\to\infty} \int_0^1 U^2(x,t)dx = 0.$$

If in addition σ satisfies the condition $|\sigma(\xi)| \leq M\left(|\xi| + |\xi|^r\right)$ for some $M > 0$ and r, $1 < r < 2$, then for any $\eta \in C^1[0,1]$ with $\eta(0) = \eta(1) = 0$,

$$\lim_{t\to\infty} \int_0^1 \frac{\partial U(x,t)}{\partial t}\eta(x)dx = 0.$$

This theorem extends to a result on approach to a steady state.

2.5.3 Rate of asymptotic behavior

MacCamy [345] has proved that the solution of the initial-boundary value problem (2.21), (2.59), and (2.60) has a power-like decay-rate:

$$\left(\sup_{x\in[0,1]} |U(x,t)|\right)^2 = O(t^{-N})$$

and

$$\int_0^1 \left(\frac{\partial U(x,t)}{\partial t}\right)^2 dx = O(t^{-N}),$$

where N is the integer in (2.61).

2.6 Comments and bibliographical notes

Mathematical models describing electromagnetic processes and many related diffusion phenomena are simulated by Maxwell's system of partial differential equations and Maxwell's-type systems.

Due to the complexity of these real problems, it is very hard to obtain the exact solutions. So, naturally the study of existence, uniqueness, asymptotic behavior, construction of discrete analogues and other main properties of the corresponding initial, boundary, and initial-boundary value problems for these models arise.

Investigation of existence, uniqueness, asymptotic behavior of the solutions, and other properties of these and related differential, integral, and integro-differential problems are studied in many works, see, e.g., [6], [13], [47], [75], [76], [77], [107], [108], [109], [122], [123], [133], [134], [136], [148], [149], [150], [151], [155], [156], [169], [175], [176], [177], [179], [180], [187], [188], [191], [199], [204], [216], [217], [218] [219], [228], [229], [231], [232], [233], [235], [237], [257], [276], [278], [290], [295], [296], [297], [298], [299], [302], [303], [304], [305], [306], [307], [309], [310], [311], [312], [322], [327], [328], [329], [332], [333], [334],

[338], [339], [340], [344], [349], [352], [358], [363], [380], [381], [383], [384], [385], [386], [405], [414], [430], [431], [436], [437], [441], [445], [451], [461], [462], [463], [465], [466], [467], [468], [469], [483], [484], [500], and reference therein.

Such investigations for Maxwell's systems and Maxwell's-type systems are given in [62], [115], [136], [139], [140], [141], [144], [145], [148], [152], [173], [174], [181], [186], [187], [188], [195], [197], [208], [224], [225], [226], [231], [232], [247], [253], [265], [266], [267], [275], [276], [290], [297], [298], [362], [365], [367], [370], [397], [409], [427], [429], [452], [453], [454], [470], [472], [473], [475], [476], [477], [485], [486], [487], [488], [489], [490], [491], and [492] among others.

In some cases the analysis and the construction of exact solutions nevertheless are possible. In this direction various types of integro-differential equations are investigated by the symmetry analysis by using group theory, see [100], [190], and references therein.

As we already have noted many different types of practical processes are described by Maxwell's systems and Maxwell's-type systems. For example, similar system arises for describing adiabatic shearing flow. To understand the effects of the dependence of the viscosity on temperature and see if it can destabilize the flow of a fluid, Dafermos and Hsiao [121] proposed a test problem. They model an adiabatic rectilinear shearing flow in an incompressible Newtonian viscous fluid between parallel plates. They assume that the left plate at $x = 0$ is at rest and the right plate at $x = 1$ moves with constant velocity v in a direction perpendicular to x. Between the plates, the flow is uniform and parallel to the plates. The balance laws of mass, momentum, and energy, assuming density and specific heat normalized to unity, can be written as:

$$\frac{\partial V}{\partial t} = \frac{\partial \sigma}{\partial x}, \quad (x,\, t) \in Q,$$

$$\frac{\partial \vartheta}{\partial t} = \sigma \frac{\partial V}{\partial x}, \quad (x,\, t) \in Q, \tag{2.64}$$

$$V(0,t) = 0, \quad V(1,t) = v, \quad t \geq 0,$$

$$V(x,0) = V_0(x), \quad \vartheta(x,0) = \vartheta_0(x), \quad x \in [0,\, 1],$$

where $V = V(x,t)$ is the velocity, σ is the shear stress, and $\vartheta = \vartheta(x,t)$ is the temperature. To close the system, the authors assume that the fluid is linearly viscous, i.e.,

$$\sigma = \mu(\vartheta)\frac{\partial V}{\partial x}. \tag{2.65}$$

Note that $V(x,t) = vx$, $\vartheta(x,t) = h(t)$ where $h(t)$ is determined by

$$\int_a^{h(t)} \frac{d\tau}{\mu(\tau)} = t$$

is a solution of (2.64). In other words, if the flow begins with a uniform shear profile it stays that way. Much of the analysis of (2.64), (2.65) in the literature concentrates understanding the longtime behavior of the flow when the initial velocity is close to $V_0(x) = vx$ and the initial temperature ϑ_0 is almost constant.

Various models have been proposed for the explanation of the existence of shear layers [158]. Much of the existing literature study the simple model (2.64) or closely related ones. The motivation is to gain a better understanding of the localization of the temperature and the formation of shear layers.

Detailed mathematical analysis of (2.64), (2.65) type systems has been carried out in [60], [94], [95], [99], [121], [139], [145], [148], [152], [158], [168], [173], [174], [276], [426], [429], [452], [453], [454], [475], [476], [477], and in a number of other works.

In mathematical modeling of many practical processes, systems with different order than the above-mentioned type equations appear. The systems of ordinary and partial differential equations are considered very often as well. For example, in [56] a nonlinear system consisting of three different types of differential equations: a first order ordinary differential equation, a parabolic and an elliptic partial differential equations is considered. Such a system may occur, e.g., as a generalization of a model describing fluid flow in porous media.

As we have noticed in the introduction one of the important hyperbolic integro-differential model is Kirchhoff equation [286]. Many works are devoted to investigation of hyperbolic type integro-differential equations, especially to Kirchhoff and its type equations, see [24], [36], [37], [55], [65], [130], [306], [309], [327], [328], [349], [355], [356], [357], [373], [380], [383], [390], [394], [395], [408], [471] and references therein

Many features of solutions of nonlinear integro-differential equations have been studied in different contexts. Very often much effort has been devoted to the study of this type models as abstract equations in Banach or Hilbert spaces [51], [52], [119], [175], [289], [327], [328].

Cauchy problem for nonlinear nonstationary abstract equations, including integro-differential equations, are investigated in many scientific works, for example, [13], [14], [44], [51], [52], [116], [119], [165], [170], [171], [175], [203], [298], [327], [328], [338], [339], [428], [455] and references therein. Similar problems were extensively studied by many authors using monotone operator techniques, see, e.g., [13], [77], [116], [175], [262], [264], [327], [364], [414], [428], [431], [455], analytic semigroup theory, see, e.g., [44], [203], [338] or classical Picard iteration, see, e.g. [87], [175]. Very often Galerkin's method is used for studying existence properties for this type of evolution integro-differential equations, see, e.g., [47], [175], [327] and [328].

Note also the huge contribution of integral and singular integral equations for mathematical modeling of real processes. Many scientific works are dedicated to their investigations too [10], [64], [91], [114], [117], [135], [159], [184], [192], [287], [291], [293], [294], [326], [330], [347], [361], [372], [396], [458], [460], [505]. The results obtained in these investigations play significant role in studying integro-differential models.

Finally note that for investigation of existence, uniqueness and asymptotic behavior of the considered initial-boundary value problems we often use Sobolev spaces and their properties, see [7], [136], [175], [327] and references therein. In this book the L_p and C^k spaces are usually used as well.

We now give detailed bibliographic references and short comments for specific sections of Chapter 2.

2.1.1 General statement of diffusion process

General statement of diffusion process of penetration of a magnetic field into a substance is given in many scientific works, see [92] [136], [139], [145], [148], [300], [359], [360], [399], [419], [473] and references therein.

2.1.2 A reduction of system of the nonlinear differential equations to the integro-differential model (Model I)

The reduction of the diffusion process of penetration of a magnetic field into a substance to the integro-differential (1.6), (1.7), (1.8) and (1.10) models, as we have already mentioned at first are done in [187], [188]. After this such question was discussed in some works [139], [145], [148], [276] as well.

2.2 On the averaged Model II

The reduction of the diffusion process of penetration of a

magnetic field into a substance to the integro-differential averaged (1.9) and (1.11) models, as we have already mentioned at first are done in [306]. After this such question was also discussed in some works [145] and [276].

2.3 Mathematical Model III

Mathematical model for one-dimensional heat flow in materials with memory are given in many scientific works, see [109], [196], [344], [345], [346] and references therein.

2.4.1 Existence and uniqueness of the solutions

The existence and uniqueness of the solutions of the initial-boundary value problems for (2.22) type integro-differential equations are studied in the works, see, for example, [136], [138], [139], [140], [141], [145], [186], [187] [188], [218], [219], [238], [240], [248], [302], [303], [304], [305], [306], [307], [322], [333], [334] and references therein.

In [137], [139], [140], [141], [145], [219], [304], [305], [306], [307] and in a number of other works, some investigations for multidimensional integro-differential scalar models of the same type are given.

The solvability and uniqueness property for higher order integro-differential equations were studied in [305], [306] and [307].

Note that even studying one-dimensional (2.22) integro differential type systems with two-component vector of the magnetic field have its complexities. Its investigations are continuing, see, for example, [223], [247], [253], [245] and [254].

The problems with different type of boundary and initial conditions are studied for these type models in the works [49], [50], [334].

Except classical initial-boundary value problems, for (2.22) type equations the problems having time-periodic boundary conditions (see [334]) as well as terminal value problems (see [49] and [50]) are discussed.

The existence and uniqueness of solutions of the (2.23) type averaged integro-differential models are studied in the works [145], [217], [219], [246]. For multidimensional (2.23) and more general type equations such investigations are done in [145].

Note that even studying one-dimensional integro-differential (2.23) type averaged systems with two-component vector of the magnetic field have its complexities and its investigations are continuing intensively, see, for example, [223], [247], [253], [257] and [261].

Let us also note that such investigations for more general models of (2.22) and (2.23) type are given in the works [139], [145], [219], [305], [306], [322].

Finally note that existence of solutions of initial-boundary value problems for (2.22) and (2.23) type integro-differential equations and systems, in the above-mentioned papers as a rule are established for the following nonlinearities:

$$a(S) = (1+S)^p, \quad -1/2 < p < 0, \quad 0 < p \leq 1;$$

$$a_0 \leq a(S) \leq a_1, \quad 4\int_0^\infty S\,|a'(S)|^2\,dS < \inf a^2(S);$$

or

$$a_0 S^p \leq a(S) \leq a_1 S^p + a_2, \quad 0 < p \leq 1;$$

where a_0, a_1 and a_2 are positive constants.

2.4.2 Asymptotic behavior of the solutions as time tends to infinity

The asymptotic behavior of the solutions as time tends to infinity of the solutions of first kind initial-boundary value problem for (2.22) type integro-differential equations are studied in: [145], [146], [147], [217], [218], [219], [227], [228], [235], [237], [238], [239], [240], [243], [244] and in a number of other works as well.

Large time behavior of the solution to an initial-boundary value problem with mixed boundary conditions for a (2.22) type integro-differential equation are studied in [248], [252] and [280].

Let us note that such investigations for one-dimensional integro-differential (2.22) type systems with two-component magnetic field are also given in the works [223], [229], [231], [232], [241], [245], [247], [253], [254], [276], [277], [282] and [283].

The asymptotic behavior of the solutions as time tends to infinity of the solutions of first kind initial-boundary value problem for (2.23) type averaged equations are studied in the works [31], [32], [145], [217], [219], [236], [237], [239], [242], [243], [249], [251] and [256].

Large time behavior of the solutions to an initial-boundary value problem with mixed boundary conditions for a (2.23) type averaged integro-differential equation are studied in [246] and [280].

Let us note that such investigations for one-dimensional integro-differential (2.23) type averaged systems with two-component magnetic field are also given in: [33], [34], [223], [231], [233], [234], [241], [247], [253], [257], [261], [276], [278] and [282].

Such investigations for more general integro-differential models of (2.22) and (2.23) type are given in the works [145], [219] and [276]. In [28], [29] and [35] such investigations are also done for (2.22) type equations and (2.23) type systems, respectively, with source terms.

2.4.3 Rate of the asymptotic behavior of solutions of Model I

Rate of asymptotic behavior of the solutions as time tends to infinity of the (2.22) type models are studied in the works [145], [146], [147], [218], [219], [227], [228], [235], [237], [238], [239], [243] and [244].

Let us note that such investigations for one-dimensional (2.22) type integro-differential systems with two-component magnetic field are also given in the works [223], [229], [231], [232], [241], [245], [247], [253], [254], [276], [277] and [283].

Large time behavior of the solution to an initial-boundary value problem with mixed boundary conditions for a (2.22) type integro-differential equation are discussed in [280].

We note also that the exponential stabilization in the norm of the space L_2 of the solutions of the initial-boundary value problems for (2.22) type equations with whole homogenous boundary conditions are proved. Such results for the following nonhomogeneous boundary condition on one side of lateral boundary:

$$U(0,t) = 0, \qquad U(1,t) = \psi, \tag{2.66}$$

where $\psi = const > 0$ are also established. These two problems for the general case of nonlinearity $a = a(S) \geq const > 0$ for (2.22) equations in works [146], [217], [218], [228], [235], [237] and [239] while (2.22) type one-dimensional integro-differential systems with two-component magnetic fields are studied in the following works: [223], [238], [254], [276] and [277]. In the norm of the space C^1 such exponential behavior for homogeneous boundary conditions, are given in [227], [237], [240], [243], [245], [277] and [283] for the cases:

$$a(S) = (1+S)^p, \quad -1/2 < p < 0, \qquad 0 < p \leq 1. \tag{2.67}$$

As we already have noticed, first kind initial-boundary value problems with nonhomogeneous boundary condition on one side of the lateral boundary, that is, the problem of (2.66) boundary conditions with nonzero condition on right boundary are also discussed. For (2.22) kind integro-differential equations under the above-mentioned (2.67) kind nonlinearities the following power-like stabilizations in the norm of the space C^1 are obtained (see [147], [218], [219], [227], [231], [240], [241] and [243]):

$$\frac{\partial U}{\partial x} = \psi + O(t^{-1-p}), \quad \frac{\partial U}{\partial t} = O(t^{-1}). \tag{2.68}$$

Let us note that such results for one-dimensional (2.22) type integro-differential systems with two-component magnetic fields and first kind initial-boundary value problems with nonhomogeneous boundary conditions on one side of lateral boundary analogous power-like (2.68) kind stabilization are also proved in [223], [229], [232], [245], [254], [276] and [283].

2.4.4 Rate of the asymptotic behavior of solutions of Model II

Rate of asymptotic behavior of the solutions as time tends to infinity of the first kind initial-boundary value problems for (2.23) type averaged integro-differential equations are studied in the works [31], [32], [145], [217], [219], [236], [237], [239], [242], [243], [246], [256]. In [28], [35] such investigation are done for (2.23) type equations with source term.

Large time behavior of the solution to an initial-boundary value problem with mixed boundary conditions for a (2.23) type averaged integro-differential equations are studied in [246] and [280].

Let us note that such investigations for one-dimensional (2.23) type averaged integro-differential systems with two-component magnetic field are also given in the works [33], [34], [223], [231], [233], [234], [241], [247], [253], [257], [261], [276] and [278]. In [29] such investigation are done for (2.23) type systems with source terms.

We note also that the exponential stabilization in the norm of the space L_2, of the solutions of the first kind initial-boundary value problems for (2.23) type averaged equations with homogenous boundary conditions on whole boundary are proved. Such results for nonhomogeneous boundary condition on one side of lateral boundary (2.66) are also established. These two problems for the general case of nonlinearity $a = a(S) \geq const > 0$ for (2.23) equations were studied in [236], [237] and [239] while (2.23) type one-dimensional integro-differential systems with two-component magnetic fields were studied in the following works [234] and [276]. In the norm of the space C^1 such exponential rate behavior for homogeneous boundary conditions, are also given in [223], [237], [241], [242], [243], [256], [257], [261] and [278] for the cases:

$$a(S) = (1 + S)^p, \quad -1/2 < p < 0, \quad p > 0. \tag{2.69}$$

Problems with nonhomogeneous boundary condition (2.66) on one side of the lateral boundary are also discussed. For (2.23) kind integro-differential averaged equations under the (2.69) nonlinearities the above-mentioned (2.68) power-like stabilizations in the norm of the space C^1 are also obtained [219], [237] and [239].

Let us note that such results for one-dimensional (2.23) type integro-differential systems with two-component magnetic fields and first kind initial-boundary value problems with nonhomogeneous boundary conditions on one side of lateral boundary analogous power-like stabilization (2.68) are also proved in [231], [233], [234], [257], [276] and [278].

Sometimes exponential stabilization are also proved for this type problem in case of (2.23) averaged equation (see [31], [32], [236] and [237]) and one-dimensional (2.23) type integro-differential systems with two-component magnetic fields (see [33], [34], [223] and [241]).

2.5 Some features of Model III

Some mathematical features of mathematical model (2.21) for one-dimensional heat flow in materials with memory are given in [332], [345], [438] and in a number of other works.

At last let us note that asymptotic behavior of solutions of the considered problems is established by numerical experiments too. Results of these experiments are described and the appropriate graphical illustrations are given in Chapter 4 of this book.

Chapter 3

Approximate Solutions of the Integro-Differential Models

Abstract

This chapter consists of nine sections. In this chapter algorithms for approximate solutions for the previously described integro-differential models are considered. The first five sections are devoted to the construction and investigation of semi-discrete scheme in space and finite difference schemes for the models investigated in the book. Theorems of stability and convergence are proved. The next three sections are dedicated to the investigation of the Galerkin's method for all three models. Convergence of Galerkin's method and error estimates are given. In the last section the bibliographical overview of the issues considered in the third chapter is done.

Key words: Semi-discrete scheme, finite-difference scheme, Galerkin's method, finite element approximation, error estimate, stability and convergence.

Numerical Solutions of Three Classes of Nonlinear Parabolic Integro-Differential Equation.
http://dx.doi.org/10.1016/B978-0-12-804628-9.50003-X

3.1 Semi-discrete scheme for Model I

In the rectangle $\overline{Q}_T$ we discuss semi-discrete scheme with respect to space variable for approximating the solution of the following nonlinear integro-differential problem:

$$\frac{\partial U}{\partial t} - \frac{\partial}{\partial x}\left\{\left[1+\int_0^t \left(\frac{\partial U}{\partial x}\right)^2 d\tau\right]\frac{\partial U}{\partial x}\right\} \tag{3.1}$$

$$= f(x,t), \quad (x,\, t) \in \overline{Q}_T,$$

$$U(0,t) = U(1,t) = 0, \quad t \in [0,\, T], \tag{3.2}$$

$$U(x,0) = U_0(x), \quad x \in [0,\, 1]. \tag{3.3}$$

Here $f = f(x,t)$ and $U_0 = U_0(x)$ are given functions of their arguments.

In order to describe the semi-discretization of problem (3.1)-(3.3) we introduce a net whose mesh points are denoted by $x_i = ih$, $i = 0, 1, \ldots, M$, with $h = 1/M$. The boundaries are specified by $i = 0$ and $i = M$. In this section the semi-discrete approximation at (x_i, t) is designated $u_i = u_i(t)$. The exact solution to problem (3.1)-(3.3) at (x_i, t), denoted by $U_i = U_i(t)$, is assumed to exist and be smooth enough. From the boundary conditions (3.2) we have $u_0(t) = u_M(t) = 0$. At other points $i = 1, 2, \ldots, M-1$, the integro-differential equation will be replaced by approximating the space derivatives by a forward and backward differences. We will use the following known notations [411]:

$$u_{x,i}(t) = \frac{u_{i+1}(t) - u_i(t)}{h}, \quad u_{\bar{x},i}(t) = \frac{u_i(t) - u_{i-1}(t)}{h}.$$

Note that the values $u_i(0)$, $i = 1, 2, \dots, M-1$ can be computed from the initial condition (3.3), i.e., $u_i(0) = U_{0,i}, \quad i = 1, 2, \dots, M-1$.

Let us correspond to problem (3.1)-(3.3) the semi-discrete scheme:

$$\frac{du_i}{dt} - \left\{ \left[1 + \int_0^t \left(u_{\bar{x},\,i}\right)^2 d\tau \right] u_{\bar{x},\,i} \right\}_x = f(x_i, t), \tag{3.4}$$

$$i = 1, 2, \dots, M-1,$$

$$u_0(t) = u_M(t) = 0, \tag{3.5}$$

$$u_i(0) = U_{0,i}, \quad i = 0, 1, \dots, M. \tag{3.6}$$

So, we obtained a Cauchy problem (3.4)-(3.6) for nonlinear system of ordinary integro-differential equations.

Introduce inner products and norms [411]:

$$(u, v)_h = \sum_{i=1}^{M-1} u_i v_i h, \quad (u, v]_h = \sum_{i=1}^{M} u_i v_i h,$$

$$||u||_h = (u, u)_h^{1/2}, \quad ||u]|_h = (u, u]_h^{1/2}.$$

Multiplying equations (3.4) by $u(t) = (u_1(t), u_2(t), \dots, u_{M-1}(t))$, using the discrete analogue of the integration by parts and the discrete analogue of Poincaré's inequality [411]

$$\|u(t)\|_h \le \|u_{\bar{x}}(t)]|_h,$$

we get

$$\frac{1}{2}\frac{d}{dt}\|u(t)\|_h^2 + \|u_{\bar{x}}(t)]|_h^2 \le (f(t), u(t))$$

$$\le \frac{1}{2}\|f(t)\|_h^2 + \frac{1}{2}\|u(t)\|_h^2 \le \frac{1}{2}\|f(t)\|_h^2 + \frac{1}{2}\|u_{\bar{x}}(t)]|_h^2,$$

where $f(t) = (f_1(t), f_2(t), \ldots, f_{M-1}(t))$, $f_i(t) = f(x_i, t)$. So, we have

$$\|u(t)\|_h^2 + \int_0^t \|u_{\bar{x}}]|_h^2 d\tau < C, \tag{3.7}$$

where C here and below in this section denotes a positive constant which does not depend on h.

The a-priori estimate (3.7) guarantees the global solvability of problem (3.4)-(3.6) (see, for example, [133], [192], [299], [326], [327]).

The principal aim of the present section is the proof of the following statement.

Theorem 3.1 *If problem (3.1)-(3.3) has a sufficiently smooth solution $U = U(x,t)$, then the solution $u = u(t) = (u_1(t), \ldots, u_{M-1}(t))$ of problem (3.4)-(3.6) tends to $U = U(t) = (U_1(t), \ldots, U_{M-1}(t))$ as $h \to 0$ and the following estimate is true*

$$\|u(t) - U(t)\|_h \leq Ch. \tag{3.8}$$

Proof. For the exact solution $U = U(x,t)$ we have:

$$\frac{dU_i}{dt} - \left\{\left[1 + \int_0^t (U_{\bar{x},i})^2 d\tau\right] U_{\bar{x},i}\right\}_x = f(x_i,t) - \psi_i(t), \tag{3.9}$$

$$i = 1, 2, \ldots, M-1,$$

$$U_0(t) = U_M(t) = 0, \tag{3.10}$$

$$U_i(0) = U_{0,i}, \quad i = 0, 1, \ldots, M, \tag{3.11}$$

where

$$\psi_i(t) = O(h).$$

Let $z_i(t) = u_i(t) - U_i(t)$ be the difference between the approximate and exact solutions. From (3.4)-(3.6) and (3.9)-(3.11) we have:

$$\begin{aligned}\frac{dz_i}{dt} &- \left\{\left[1+\int_0^t (u_{\bar{x},i})^2 d\tau\right] u_{\bar{x},i}\right.\\ &- \left.\left[1+\int_0^t (U_{\bar{x},i})^2 d\tau\right] U_{\bar{x},i}\right\}_x = \psi_i(t),\end{aligned} \tag{3.12}$$

$$z_0(t) = z_M(t) = 0, \tag{3.13}$$

$$z_i(0) = 0. \tag{3.14}$$

Multiplying (3.12) by $z(t) = (z_1(t), z_2(t), \ldots, z_{M-1}(t))$, using (3.13) and the discrete analogue of the integration by parts [411], we get

$$\begin{aligned}&\frac{1}{2}\frac{d}{dt}\|z\|_h^2 + \sum_{i=1}^{M} \left\{\left[1+\int_0^t (u_{\bar{x},i})^2 d\tau\right] u_{\bar{x},i}\right.\\ &- \left.\left[1+\int_0^t (U_{\bar{x},i})^2 d\tau\right] U_{\bar{x},i}\right\} z_{\bar{x},i}h = \sum_{i=1}^{M-1} \psi_i z_i h.\end{aligned} \tag{3.15}$$

Note that

$$\left\{\left[1+\int_0^t (u_{\bar{x},i})^2 d\tau\right] u_{\bar{x},i} - \left[1+\int_0^t (U_{\bar{x},i})^2 d\tau\right] U_{\bar{x},i}\right\} (u_{\bar{x},i} - U_{\bar{x},i})$$

$$= (z_{\bar{x},i})^2 + \left[\int_0^t (u_{\bar{x},i})^2 d\tau u_{\bar{x},i} - \int_0^t (U_{\bar{x},i})^2 d\tau U_{\bar{x},i}\right] (u_{\bar{x},i} - U_{\bar{x},i})$$

Therefore

$$\left\{\left[1+\int_0^t (u_{\bar{x},i})^2 d\tau\right] u_{\bar{x},i} - \left[1+\int_0^t (U_{\bar{x},i})^2 d\tau\right] U_{\bar{x},i}\right\} (u_{\bar{x},i} - U_{\bar{x},i})$$

$$= (z_{\bar{x},i})^2 + \frac{1}{2}\left[\int_0^t (u_{\bar{x},i})^2 d\tau + \int_0^t (U_{\bar{x},i})^2 d\tau\right] (u_{\bar{x},i} - U_{\bar{x},i})^2$$

$$+\frac{1}{2}\left[\int_0^t (u_{\bar{x},i})^2 d\tau - \int_0^t (U_{\bar{x},i})^2 d\tau\right] \left[(u_{\bar{x},i})^2 - (U_{\bar{x},i})^2\right]$$

$$\geq \frac{1}{2}\left[\int_0^t (u_{\bar{x},i})^2 d\tau - \int_0^t (U_{\bar{x},\imath})^2 d\tau\right] \left[(u_{\bar{x},i})^2 - (U_{\bar{x},i})^2\right] . \tag{3.16}$$

Using (3.15) and (3.16), we have

$$\frac{1}{2}\frac{d}{dt}\|z\|_h^2 + \frac{1}{2}\sum_{i=1}^{M} \frac{d}{dt}\varphi_i^2(t) h \leq \sum_{i=1}^{M-1} \psi_i z_i h, \tag{3.17}$$

where

$$\varphi_i(t) = \int_0^t \left[(u_{\bar{x},i})^2 - (U_{\bar{x},i})^2\right] d\tau.$$

Integration and using (3.14) in (3.17), we get

$$\|z(t)\|_h^2 \leq \int_0^t \|z(\tau)\|_h^2 d\tau + \int_0^t \|\psi(\tau)\|_h^2 d\tau. \tag{3.18}$$

Using Grönwall's lemma from (3.18) we get (3.8), and Theorem 3.1 has been proved.

Note that according to the scheme of proving convergence in Theorem 3.1, the uniqueness of the solution of the semi-discrete scheme (3.4)-(3.6) can be proved. In particular, assuming existence of two solutions $u(t)$ and $\bar{u}(t)$ of the scheme (3.4)-(3.6), for the difference $\bar{z}(t) = u(t) - \bar{u}(t)$ we get

$$\|z(t)\|_h^2 \le \int_0^t \|z(\tau)\|_h^2 d\tau.$$

So, using Grönwall's lemma again, we have $\bar{z} \equiv 0$, i.e., $u(t) \equiv \bar{u}(t)$.

3.2 Finite difference scheme for Model I

In order to describe the finite difference method we introduce a net whose mesh points are denoted by $(x_i, t_j) = (ih, j\tau)$, where $i = 0, 1, ..., M$; $j = 0, 1, ..., N$ with $h = 1/M$, $\tau = T/N$. The initial line is denoted by $j = 0$. The discrete approximation at (x_i, t_j) is denoted by u_i^j and the exact solution to problem (3.1)-(3.3) at those points by U_i^j. We will use the following known notations [411]:

$$u_{t,i}^j = \frac{u_i^{j+1} - u_i^j}{\tau}, \quad u_{\bar{t},i}^j = u_{t,i}^{j-1} = \frac{u_i^j - u_i^{j-1}}{\tau}.$$

Let us correspond to problem (3.1)-(3.3) the difference scheme:

$$\frac{u_i^{j+1} - u_i^j}{\tau} - \left\{ \left[1 + \tau \sum_{k=1}^{j+1} (u_{\bar{x},i}^k)^2 \right] u_{\bar{x},i}^{j+1} \right\}_x = f_i^j, \tag{3.19}$$

$$i = 1, 2, ..., M-1; \quad j = 0, 1, ..., N-1,$$

$$u_0^j = u_M^j = 0, \quad j = 0, 1, ..., N, \tag{3.20}$$

$$u_i^0 = U_{0,i}, \quad i = 0, 1, ..., M. \tag{3.21}$$

Multiplying equations (3.19) by $u^j = (u_1^{j+1}, u_2^{j+1}, \dots, u_{M-1}^{j+1})$ using (3.20), (3.21) and the discrete analogue of integration by parts it is not difficult to prove the inequality

$$\|u^n\|_h^2 + \tau \sum_{j=1}^{n} \|u_{\bar{x}}^j]|_h^2 \leq C, \quad n = 1, 2, ..., N, \tag{3.22}$$

where C here and below in this section is a positive constant independent of τ and h.

The a-priori estimate (3.22) guarantees the stability of the scheme (3.19)-(3.21). From (3.22) it is not difficult to get solvability of the system of nonlinear equations (3.19)-(3.21) as well (see, for example, [327]).

The principal aim of the present section is the proof of the following statement.

Theorem 3.2 *If problem (3.1)-(3.3) has a sufficiently smooth solution $U = U(x,t)$, then the solution $u^j = (u_1^j, u_2^j, \dots, u_{M-1}^j)$, $j = 1, 2, \dots, N$ of the difference scheme (3.19)-(3.21) tends to $U^j = (U_1^j, U_2^j, \dots, U_{M-1}^j)$, $j = 1, 2, \dots, N$ as $\tau \to 0$, $h \to 0$, and the following estimate holds*

$$\|u^j - U^j\|_h \leq C(\tau + h), \ j = 1, 2, \dots, N. \tag{3.23}$$

Proof. For the exact solution $U = U(x,t)$ of problem (3.1)-(3.3) we have:

$$\frac{U_i^{j+1} - U_i^j}{\tau} - \left\{ \left[1 + \tau \sum_{k=1}^{j+1} (U_{\bar{x},i}^k)^2 \right] U_{\bar{x},i}^{j+1} \right\}_x = f_i^j - \psi_i^j, \tag{3.24}$$

$$U_0^j = U_M^j = 0, \tag{3.25}$$

$$U_i^0 = U_{0,i}, \tag{3.26}$$

where

$$\psi_i^j = O(\tau + h).$$

Solving (3.19)-(3.21) instead of problem (3.1)-(3.3) we have the error $z_i^j = u_i^j - U_i^j$. From (3.19)-(3.21) and (3.24) - (3.26) we get:

$$\begin{aligned} \frac{z_i^{j+1} - z_i^j}{\tau} &- \left\{ \left[1 + \tau \sum_{k=1}^{j+1} (u_{\bar{x},i}^k)^2 \right] u_{\bar{x},i}^{j+1} \right. \\ &- \left. \left[1 + \tau \sum_{k=1}^{j+1} (U_{\bar{x},i}^k)^2 \right] U_{\bar{x},i}^{j+1} \right\}_x = \psi_i^j, \end{aligned} \tag{3.27}$$

$$z_0^j = z_M^j = 0, \tag{3.28}$$

$$z_i^0 = 0. \tag{3.29}$$

Multiplying (3.27) by $z^{j+1} = (z_1^{j+1}, z_2^{j+1}, ..., z_{M-1}^{j+1})$, using (3.28), and the discrete analogue of integration by parts we get

$$\begin{aligned} \|z^{j+1}\|_h^2 &\quad (z^{j+1}, z^j)_h \\ &+ \tau h \sum_{i=1}^{M} \left\{ \left[1 + \tau \sum_{k=1}^{j+1} (u_{x,i}^k)^2 \right] u_{\bar{x},i}^{j+1} \right. \\ &- \left. \left[1 + \tau \sum_{k=1}^{j+1} (U_{\bar{x},i}^k)^2 \right] U_{\bar{x},i}^{j+1} \right\} z_{\bar{x},i}^{j+1} \\ &= \tau (\psi^j, z^{j+1})_h. \end{aligned} \tag{3.30}$$

Taking into account the relations:

$$(z^{j+1}, z^j)_h = \frac{1}{2}\|z^{j+1}\|_h^2 + \frac{1}{2}\|z^j\|_h^2 - \frac{1}{2}\|z^{j+1} - z^j\|_h^2,$$

$$\left[(u^k_{\bar{x},i})^2 u^{j+1}_{\bar{x},i} - (U^k_{\bar{x},i})^2 U^{j+1}_{\bar{x},i}\right]\left(u^{j+1}_{\bar{x},i} - U^{j+1}_{\bar{x},i}\right)$$

$$\geq \frac{1}{2}\left[(u^k_{\bar{x},i})^2 - (U^k_{\bar{x},i})^2\right]\left[(u^{j+1}_{\bar{x},i})^2 - (U^{j+1}_{\bar{x},i})^2\right],$$

from (3.30) we have

$$\|z^{j+1}\|_h^2 + \frac{1}{2}\|z^{j+1} - z^j\|_h^2$$

$$-\frac{1}{2}\|z^{j+1}\|_h^2 - \frac{1}{2}\|z^j\|_h^2 + \tau\|z^{j+1}_{\bar{x}}]|_h^2$$

$$+\frac{\tau^2 h}{2}\sum_{i=1}^{M-1}\sum_{k=1}^{j+1}\left[(u^k_{\bar{x},i})^2 - (U^k_{\bar{x},i})^2\right]\left[(u^{j+1}_{\bar{x},i})^2 - (U^{j+1}_{\bar{x},i})^2\right] \tag{3.31}$$

$$\leq \frac{\tau}{2\varepsilon}\|\psi^j\|_h^2 + 2\varepsilon\tau\|z^{j+1}\|_h^2,$$

$$\forall\varepsilon > 0, \quad j = 0, 1, ..., N-1.$$

Introduce the notations

$$\xi^j_i = \tau\sum_{k=1}^{j}\left[(u^k_{\bar{x},i})^2 - (U^k_{\bar{x},i})^2\right],$$

then

$$\xi^j_{i,t} = \left[(u^{j+1}_{\bar{x},i})^2 - (U^{j+1}_{\bar{x},i})^2\right].$$

So, from (3.31) we get

$$\|z^{j+1}\|_h^2 - \|z^j\|_h^2 + \tau^2\|z^{j+1}_{\bar{t}}\|_h^2 + \tau\|z^{j+1}_{\bar{x}}]|_h^2$$

$$+\tau^2\|\xi^j_t\|_h^2 + \tau\left(\xi^j, \xi^j_t\right)_h \leq \frac{\tau}{\varepsilon}\|\psi^j\|_h^2 + 4\varepsilon\tau\|z^{j+1}\|_h^2. \tag{3.32}$$

Using (3.29), the discrete analogue of Poincaré's inequality [411]

$$\|z^{j+1}\|_h^2 \le \frac{1}{8}\|z_{\bar{x}}^{j+1}]|_h^2$$

and the relation

$$\tau\left(\xi^j, \xi_t^j\right)_h = \frac{1}{2}\|\xi^{j+1}\|_h^2 - \frac{1}{2}\|\xi^j\|_h^2 - \frac{\tau^2}{2}\|\xi_t^j\|_h^2,$$

we have from (3.32)

$$\|z^n\|_h^2 + \tau^2\sum_{j=0}^{n-1}\|z_{\bar{t}}^{j+1}\|_h^2 + \frac{\tau}{2}\sum_{j=0}^{n-1}\|z_{\bar{x}}^{j+1}]|_h^2$$

$$+\frac{\tau^2}{2}\sum_{j=0}^{n-1}\|\xi_t^j\|_h^2 + \frac{1}{2}\|\xi^n\|_h^2 \le C\sum_{j=0}^{n-1}\|\psi^j\|_h^2\tau, \qquad (3.33)$$

$$n = 1, 2, \ldots, N.$$

From (3.33) we get (3.23) and thus Theorem 3.2 has been proved.

Note that according to the scheme of proving convergence in Theorem 3.2, the uniqueness of the solution of the scheme (3.19)-(3.21) can be proved. In particular, assuming existence of two solutions u and $\bar{u}$ of the scheme (3.19)-(3.21), for the difference $\bar{z} = u - \bar{u}$ we get $\|\bar{z}^n\|_h^2 \le 0, \quad n = 1, 2, ..., N$.

So, $\bar{z} \equiv 0$, i.e., $u \equiv \bar{u}$.

Note that for solving the difference scheme (3.19)-(3.21) we use Newton's iterative process and we have carried out various numerical experiments. These experiments demonstrate agreement with the theoretical results (see section 4.1 of Chapter 4).

3.3 Semi-discrete scheme for Model II

In the rectangle $\overline{Q}_T$ consider the following initial-boundary value problem:

$$\frac{\partial U}{\partial t} - \left[1 + \int_0^t \int_0^1 \left(\frac{\partial U}{\partial x}\right)^2 dx d\tau\right] \frac{\partial^2 U}{\partial x^2} \tag{3.34}$$

$$= f(x,t), \quad (x,\, t) \in \overline{Q}_T,$$

$$U(0,t) = U(1,t) = 0, \quad t \in [0,\, T], \tag{3.35}$$

$$U(x,0) = U_0(x), \quad x \in [0,\, 1]. \tag{3.36}$$

Let us construct the semi-discrete problem corresponding to (3.34)-(3.36):

$$\frac{du_i}{dt} - \left[1 + h \sum_{l=1}^{M} \int_0^t (u_{\bar{x},l})^2 \, d\tau\right] u_{\bar{x}x,i} = f(x_i,t), \tag{3.37}$$

$$i = 1, 2, \ldots, M-1,$$

$$u_0(t) = u_M(t) = 0, \tag{3.38}$$

$$u_i(0) = U_{0,i}, \quad i = 0, 1, \ldots, M. \tag{3.39}$$

So, we obtained a Cauchy problem (3.37)-(3.39) for nonlinear system of ordinary integro-differential equations.

Similar to the way we obtained (3.7), we can show that

$$\|u(t)\|_h^2 + \int_0^t \|u_{\bar{x}}]|_h^2 d\tau \le C. \tag{3.40}$$

Here and below in the investigation of (3.37)-(3.39), C denotes a positive constant independent of h.

The a-priori estimate (3.40) guarantees the global solvability of problem (3.37)-(3.39) (see, e.g., [133], [192], [299], [326], [327]).

The main result of this section is the following statement.

Theorem 3.3 *If problem (3.34)-(3.36) has a sufficiently smooth solution $U = U(x,t)$, then $u = u(t) = (u_1(t), u_2(t), \ldots, u_{M-1}(t))$ the solution of problem (3.37)-(3.39) tends to $U = U(t) = (U_1(t), U_2(t), \ldots, U_{M-1}(t))$ as $h \to 0$ and the following estimate holds*

$$\|u(t) - U(t)\|_h \leq Ch. \tag{3.41}$$

Proof. For the exact solution $U = U(x,t)$ we have:

$$\frac{dU_i}{dt} - \left[1 + h\sum_{l=1}^{M}\int_0^t (U_{\bar{x},l})^2 d\tau\right] U_{\bar{x}x,i} \tag{3.42}$$
$$= f(x_i,t) - \psi_i(t), \qquad i = 1,2,\ldots,M-1,$$

$$U_0(t) = U_M(t) = 0, \tag{3.43}$$

$$U_i(0) = U_{0,i}, \quad i = 0,1,\ldots,M, \tag{3.44}$$

where

$$\psi_i(t) = O(h).$$

Let $z_i(t) = u_i(t) - U_i(t)$ be the difference between the approximate and exact solutions. From (3.37)-(3.39) and (3.42)-(3.44) we have:

$$\frac{dz_i}{dt} - \left\{\left[1 + h\sum_{l=1}^{M}\int_0^t (u_{\bar{x},l})^2 d\tau\right] u_{\bar{x},i} - \left[1 + h\sum_{l=1}^{M}\int_0^t (U_{\bar{x},l})^2 d\tau\right] U_{\bar{x},i}\right\}_x = \psi_i(t), \tag{3.45}$$

$$i = 1, 2, \ldots, M-1,$$

$$z_0(t) = z_M(t) = 0, \tag{3.46}$$

$$z_i(0) = 0. \tag{3.47}$$

Multiplying (3.45) by $z(t) = (z_1(t), z_2(t), \ldots, z_{M-1}(t))$, using (3.46) and the discrete analogue of integration by parts, we get

$$\frac{1}{2}\frac{d}{dt}\|z\|_h^2 + \sum_{i=1}^{M}\left\{\left[1 + h\sum_{l=1}^{M}\int_0^t (u_{\bar{x},l})^2 d\tau\right] u_{\bar{x},i} - \left[1 + h\sum_{l=1}^{M}\int_0^t (U_{\bar{x},l})^2 d\tau\right] U_{\bar{x},i}\right\} z_{\bar{x},i}h = \sum_{i=1}^{M-1}\psi_i z_i h. \tag{3.48}$$

Note that

$$\left\{\left[1 + h\sum_{l=1}^{M}\int_0^t (u_{\bar{x},l})^2 d\tau\right] u_{\bar{x},i} - \left[1 + h\sum_{l=1}^{M}\int_0^t (U_{\bar{x},l})^2 d\tau\right] U_{\bar{x},i}\right\}(u_{\bar{x},i} - U_{\bar{x},i})$$

$$= (z_{\bar{x},i})^2 + h\sum_{l=1}^{M}\left[\int_0^t (u_{\bar{x},l})^2 d\tau u_{\bar{x},i} - \int_0^t (U_{\bar{x},l})^2 d\tau U_{\bar{x},i}\right]$$

$$\times(u_{\bar{x},i} - U_{\bar{x},i})$$

$$= (z_{\bar{x},i})^2 + \frac{1}{2}h\sum_{l=1}^{M}\left\{\left[\int_0^t (u_{\bar{x},l})^2 d\tau + \int_0^t (U_{\bar{x},l})^2 d\tau\right]\right.$$

$$\times(u_{\bar{x},i} - U_{\bar{x},i})^2 \tag{3.49}$$

$$\left. + \left[\int_0^t (u_{\bar{x},l})^2 d\tau - \int_0^t (U_{\bar{x},l})^2 d\tau\right]\left[(u_{\bar{x},i})^2 - (U_{\bar{x},i})^2\right]\right\}$$

$$\geq \frac{1}{2}h\sum_{l=1}^{M}\left[\int_0^t (u_{\bar{x},l})^2 d\tau - \int_0^t (U_{\bar{x},l})^2 d\tau\right]$$

$$\times\left[(u_{\bar{x},i})^2 - (U_{\bar{x},i})^2\right].$$

Using (3.48) and (3.49) we have

$$\frac{1}{2}\frac{d}{dt}\|z\|_h^2 + \frac{h}{2}\sum_{l=1}^{M}\left[\int_0^t (u_{\bar{x},l})^2 d\tau - \int_0^t (U_{\bar{x},l})^2 d\tau\right]$$

$$\times h\sum_{i=1}^{M}\left[(u_{\bar{x},i})^2 - (U_{\bar{x},i})^2\right] \leq \sum_{i=1}^{M-1}\psi_i z_i h. \tag{3.50}$$

Now introduce the notation

$$\varphi(t) = h\sum_{l=1}^{M}\int_0^t \left[(u_{\bar{x},l})^2 - (U_{\bar{x},l})^2\right] d\tau,$$

(3.50) becomes

$$\frac{1}{2}\frac{d}{dt}\|z\|_h^2 + \frac{1}{4}\frac{d}{dt}\varphi^2(t) \le \sum_{i=1}^{M-1} \psi_i z_i h \le \frac{1}{2}\|z\|_h^2 + \frac{1}{2}\|\psi\|_h^2, \quad (3.51)$$

or after integrating and using (3.44) in (3.51), we get

$$\|z(t)\|_h^2 \le \int_0^t \|z(\tau)\|_h^2 d\tau + \int_0^t \|\psi(\tau)\|_h^2 d\tau. \quad (3.52)$$

Using Grönwall's lemma from (3.52) we get (3.41). This proves Theorem 3.3.

Note that according to the scheme of proving convergence in Theorem 3.3, the uniqueness of the solution of the semi-discrete scheme (3.37)-(3.39) can be proved. In particular, assuming existence of two solutions $u(t)$ and $\bar{u}(t)$ of the scheme (3.37)-(3.39), for the difference $\bar{z}(t) = u(t) - \bar{u}(t)$ we get

$$\|z(t)\|_h^2 \le \int_0^t \|z(\tau)\|_h^2 d\tau.$$

So, using Grönwall's lemma again, we have $\bar{z} \equiv 0$, i.e., $u(t) \equiv \bar{u}(t)$.

3.4 Finite difference scheme for Model II

Now let us consider the following fully discrete scheme for problem (3.34)-(3.36):

$$\frac{u_i^{j+1} - u_i^j}{\tau} - \left[1 + \tau h \sum_{l=1}^{M} \sum_{k=1}^{j+1} (u_{\bar{x},l}^k)^2\right] u_{\bar{x}x,i}^{j+1} = f_i^j, \quad (3.53)$$

$$i = 1, 2, ..., M-1; \;\; j = 0, 1, ..., N-1,$$

$$u_0^j = u_M^j = 0, \quad j = 0, 1, ..., N, \tag{3.54}$$

$$u_i^0 = U_{0,i}, \quad i = 0, 1, ..., M. \tag{3.55}$$

In a similar fashion to the way we obtained (3.22), we can show that

$$\|u^n\|_h^2 + \tau \sum_{j=1}^{n} \|u_{\bar{x}}^j]|_h^2 \leq C, \quad n = 1, 2, ..., N. \tag{3.56}$$

Here and below in this section C is a positive constant independent of τ and h.

The a-priori estimate (3.56) guarantees the stability of the scheme (3.53)-(3.55). From (3.56) it is not difficult to get solvability of the system of nonlinear equations (3.53)-(3.55) as well (see, for example, [327]).

The main result of this section is the following statement.

Theorem 3.4 *If problem (3.34)-(3.36) has a sufficiently smooth solution $U = U(x,t)$ then $u^j = (u_1^j, u_2^j, \ldots, u_{M-1}^j)$, $j = 1, 2, \ldots, N$ the solution of the finite difference scheme (3.53)-(3.55) tends to the $U^j = (U_1^j, U_2^j, \ldots, U_{M-1}^j)$, $j = 1, 2, \ldots, N$ as τ and h tend to 0 and the following estimate holds:*

$$\|u^j - U^j\|_h \leq C(\tau + h), \quad j = 1, 2, \ldots, N. \tag{3.57}$$

Proof. For the exact solution $U = U(x,t)$ of problem (3.34)-(3.36) we have:

$$\frac{U_i^{j+1} - U_i^j}{\tau} - \left[1 + \tau h \sum_{l=1}^{M} \sum_{k=1}^{j+1} (U_{\bar{x},l}^k)^2\right] U_{\bar{x}x,i}^{j+1}$$
$$= f_i^j - \psi_i^j, \tag{3.58}$$

$$i = 1, 2, \ldots, M-1, \quad j = 0, 1, \ldots, N-1,$$

$$U_0^j = U_M^j = 0, \qquad j = 0, 1, \ldots, N \tag{3.59}$$

$$U_i^0 = U_{0,i}, \qquad i = 0, 1, \ldots, M, \tag{3.60}$$

where

$$\psi_i^j = O(\tau + h).$$

Solving (3.58)-(3.60) instead of problem (3.34)-(3.36) we have the error $z_i^j = u_i^j - U_i^j$. From (3.53)-(3.55) and (3.58)-(3.60) we get:

$$\frac{z_i^{j+1} - z_i^j}{\tau} - \left\{ \left[1 + \tau h \sum_{l=1}^{M} \sum_{k=1}^{j+1} (u_{\bar{x},i}^k)^2 \right] u_{\bar{x},i}^{j+1} \right.$$
$$\left. - \left[1 + \tau h \sum_{l=1}^{M} \sum_{k=1}^{j+1} (U_{\bar{x},i}^k)^2 \right] U_{\bar{x},i}^{j+1} \right\}_x = \psi_i^j, \tag{3.61}$$

$$i = 1, 2, \ldots, M-1, \quad j = 0, 1, \ldots, N-1,$$

$$z_0^j = z_M^j = 0, \tag{3.62}$$

$$z_i^0 = 0. \tag{3.63}$$

Multiplying (3.61) by $z^{j+1} = (z_1^{j+1}, z_2^{j+1}, ..., z_{M-1}^{j+1})$, using (3.62) and the discrete analogue of integration by parts we get

$$\|z^{j+1}\|_h^2 - (z^{j+1}, z^j)_h$$
$$+ \tau h \sum_{i=1}^{M} \left\{ \left[1 + \tau h \sum_{l=1}^{M} \sum_{k=1}^{j+1} (u_{\bar{x},l}^k)^2 \right] u_{\bar{x},i}^{j+1} \right.$$
$$\left. - \left[1 + \tau h \sum_{l=1}^{M} \sum_{k=1}^{j+1} (U_{\bar{x},l}^k)^2 \right] U_{\bar{x},i}^{j+1} \right\} z_{\bar{x},i}^{j+1} \tag{3.64}$$
$$= \tau (\psi^j, z^{j+1})_h.$$

Taking into account the relations:

$$(z^{j+1}, z^j)_h = \frac{1}{2}\|z^{j+1}\|_h^2 + \frac{1}{2}\|z^j\|_h^2 - \frac{1}{2}\|z^{j+1} - z^j\|_h^2,$$

$$\left[(u_{\bar{x},l}^k)^2 u_{\bar{x},i}^{j+1} - (U_{\bar{x},l}^k)^2 U_{\bar{x},i}^{j+1}\right](u_{\bar{x},i}^{j+1} - U_{\bar{x},i}^{j+1})$$

$$= (u_{\bar{x},l}^k)^2 (u_{\bar{x},i}^{j+1})^2 + (U_{\bar{x},l}^k)^2 (U_{\bar{x},i}^{j+1})^2$$

$$-(u_{\bar{x},l}^k)^2 u_{\bar{x},i}^{j+1} U_{\bar{x},i}^{j+1} - (U_{\bar{x},l}^k)^2 U_{\bar{x},i}^{j+1} u_{\bar{x},i}^{j+1}$$

$$= (u_{\bar{x},l}^k)^2 (u_{\bar{x},i}^{j+1})^2 + (U_{\bar{x},l}^k)^2 (U_{\bar{x},i}^{j+1})^2$$

$$-\left[(u_{\bar{x},l}^k)^2 + (U_{\bar{x},l}^k)^2\right] u_{\bar{x},i}^{j+1} U_{\bar{x},i}^{j+1}$$

$$\geq (u_{\bar{x},l}^k)^2 (u_{\bar{x},i}^{j+1})^2 + (U_{\bar{x},l}^k)^2 (U_{\bar{x},i}^{j+1})^2$$

$$-\frac{1}{2}\left[(u_{\bar{x},l}^k)^2 + (U_{\bar{x},l}^k)^2\right]\left[(u_{\bar{x},i}^{j+1})^2 + (U_{\bar{x},i}^{j+1})^2\right]$$

$$= \frac{1}{2}(u_{\bar{x},l}^k)^2 \left[(u_{\bar{x},i}^{j+1})^2 - (U_{\bar{x},i}^{j+1})^2\right]$$

$$-\frac{1}{2}(U_{\bar{x},l}^k)^2 \left[(u_{\bar{x},i}^{j+1})^2 - (U_{\bar{x},i}^{j+1})^2\right]$$

$$= \frac{1}{2}\left[(u_{\bar{x},l}^k)^2 - (U_{\bar{x},l}^k)^2\right]\left[(u_{\bar{x},i}^{j+1})^2 - (U_{\bar{x},i}^{j+1})^2\right],$$

from (3.64) we have

$$\|z^{j+1}\|_h^2 + \frac{1}{2}\|z^{j+1} - z^j\|_h^2 - \frac{1}{2}\|z^{j+1}\|_h^2$$

$$-\frac{1}{2}\|z^j\|_h^2 + \tau\|z_{\bar{x}}^{j+1}]|_h^2$$

$$+\frac{\tau^2 h^2}{2}\sum_{i=1}^{M}\sum_{l=1}^{M}\sum_{k=1}^{j+1}\left[(u_{\bar{x},l}^k)^2 - (U_{\bar{x},l}^k)^2\right] \tag{3.65}$$

$$\times\left[(u_{\bar{x},i}^{j+1})^2 - (U_{\bar{x},i}^{j+1})^2\right]$$

$$\leq \frac{\tau}{2\varepsilon}\|\psi^j\|_h^2 + 2\varepsilon\tau\|z^{j+1}\|_h^2,$$

$$\forall\varepsilon > 0, \quad j = 0, 1, ..., N-1.$$

Introduce the notations

$$\xi^j = \tau h\sum_{k=1}^{j}\sum_{l=1}^{M}\left[(u_{\bar{x},l}^k)^2 - (U_{\bar{x},l}^k)^2\right], \ \ \xi^0 = 0\,,$$

then

$$\xi_t^j = h\sum_{i=1}^{M}\left[(u_{\bar{x},i}^{j+1})^2 - (U_{\bar{x},i}^{j+1})^2\right].$$

So, from (3.65) we get

$$\|z^{j+1}\|_h^2 - \|z^j\|_h^2 + \tau^2\|z_t^{j+1}\|_h^2$$

$$+\tau\|z_{\bar{x}}^{j+1}]|_h^2 + \tau^2\left(\xi_t^j\right)^2 + \tau\xi^j\xi_t^j \tag{3.66}$$

$$\leq \frac{\tau}{\varepsilon}\|\psi^j\|_h^2 + 4\varepsilon\tau\|z^{j+1}\|_h^2.$$

Using (3.63) and the discrete analogue of Poincaré's inequality

$$\|z^{j+1}\|_h^2 \le \frac{1}{8}\|z_{\bar{x}}^{j+1}]|_h^2$$

and the relation

$$\tau \xi^j \xi_t^j = \frac{1}{2}\left(\xi^{j+1}\right)^2 - \frac{1}{2}\left(\xi^j\right)^2 - \frac{\tau^2}{2}\left(\xi_t^j\right)^2,$$

we have from (3.66)

$$\|z^n\|_h^2 + \tau^2 \sum_{j=0}^{n-1} \|z_{\bar{t}}^{j+1}\|_h^2 + \frac{\tau}{2}\sum_{j=0}^{n-1} \|z_{\bar{x}}^{j+1}]|_h^2$$

$$+\frac{\tau^2}{2}\sum_{j=0}^{n-1}\left(\xi_t^j\right)^2 + \frac{1}{2}\left(\xi^n\right)^2 \tag{3.67}$$

$$\le C\sum_{j=0}^{n-1} \|\psi^j\|_h^2 \tau, \quad n = 1, 2, \ldots, N.$$

From (3.67) we get (3.57) and thus Theorem 3.4 has been proved.

Note that according to the scheme of proving convergence in Theorem 3.4, the uniqueness of the solution of the scheme (3.53)-(3.55) can be proved. In particular, assuming existence of two solutions u and $\bar{u}$ of the scheme (3.53)-(3.55), for the difference $\bar{z} = u - \bar{u}$ we get $\|\bar{z}^n\|_h^2 \le 0, \quad n = 1, 2, ..., N$.

So, $\bar{z} \equiv 0$, i.e., $u \equiv \bar{u}$.

Note that for solving the difference scheme (3.53) - (3.55) we use Newton's iterative process and we have carried out various numerical experiments. These experiments demonstrate agreement with the theoretical results (see section 4.2 of Chapter 4).

3.5 Discrete analogues of Model III

In this section we consider the first type initial-boundary value problem for Model III:

$$\frac{\partial U(x,t)}{\partial t} = \int_0^t a(t-\tau)\frac{\partial}{\partial x}\sigma\left(\frac{\partial U(x,\tau)}{\partial x}\right)d\tau + f(x,t), \quad (3.68)$$

$$U(0,t) = U(1,t) = 0, \quad (3.69)$$

$$U(x,0) = U_0(x). \quad (3.70)$$

Let us describe the numerical solution of initial-boundary value problem (3.68)-(3.70) by finite difference scheme.

The boundary conditions (3.69) lead to

$$u_0^j = u_M^j = 0.$$

The difference approximation used by Neta and Igwe [378] for problem (3.68)-(3.70) is as follows:

$$\frac{\partial U(x_i,t_j)}{\partial t} = \frac{U_i^{j+1} - U_i^j}{\tau} + O(\tau), \quad (3.71)$$

$$\frac{\partial U(x_i,t_j)}{\partial x} = \frac{U_{i+1}^j - U_i^j}{h} + O(h), \quad (3.72)$$

$$\int_0^{t_j} \phi(s)ds = \sum_{\ell=1}^{j} \frac{\tau}{2}\left[\phi(t_\ell) + \phi(t_{\ell-1})\right] + O(\tau^3), \quad (3.73)$$

$$\frac{\partial}{\partial x}\sigma\left(\frac{\partial U(x_i,t_j)}{\partial x}\right) = \frac{1}{h}\left[\sigma\left(\frac{U_{i+1}^j - U_i^j}{h}\right)\right.$$
$$\left. -\sigma\left(\frac{U_i^j - U_{i-1}^j}{h}\right)\right] + O(h^2). \quad (3.74)$$

Combining (3.71), (3.73), and (3.74) with (3.68)-(3.70) for approximating the solution u, one has

$$\begin{aligned} u_i^{j+1} &= u_i^j + \tau f_i^j \\ &+\frac{\tau^2}{2}\sum_{\ell=1}^{j}\left\{\frac{a((j-\ell)\tau)}{h}\left[\sigma\left(\frac{u_{i+1}^{\ell}-u_i^{\ell}}{h}\right)\right.\right. \\ &\left.-\sigma\left(\frac{u_i^{\ell}-u_{i-1}^{\ell}}{h}\right)\right] \\ &+\frac{a((j-\ell+1)\tau)}{h}\left[\sigma\left(\frac{u_{i+1}^{\ell+1}-u_i^{\ell+1}}{h}\right)\right. \\ &\left.\left.-\sigma\left(\frac{u_i^{\ell+1}-u_{i-1}^{\ell+1}}{h}\right)\right]\right\}. \end{aligned} \tag{3.75}$$

The method is first order in time and second order in space. Note that one needs u_i^0 and u_i^1, $i = 1, 2, \ldots, M-1$ in order to use (3.75). These values can be computed from the initial condition (3.70) and from

$$\frac{\partial U(x,0)}{\partial t} = f(x,0). \tag{3.76}$$

This last relation is obtained from (3.68) upon setting $t = 0$. Therefore,

$$\begin{aligned} u_i^0 &= U_0(x_i), \\ u_i^1 &= u_i^0 + \tau f_i^0, \end{aligned} \tag{3.77}$$

and one can march forward in time.

Neta and Igwe [378] have developed a second order approximation in both time and space. This scheme has the following form

$$u_i^{j+1} = u_i^j + \tau\left(f_i^j + \frac{\tau}{2}\frac{\partial f(x_i,t_j)}{\partial t}\right)$$
$$+\tau^2\sum_{\ell=0}^{j}\zeta_\ell a(t_j - t_\ell)\frac{\sigma((u_{i+1}^\ell - u_i^\ell)/h) - \sigma((u_i^\ell - u_{i-1}^\ell)/h)}{h}, \tag{3.78}$$

where

$$\zeta_\ell = \begin{cases} 1, & \ell > 0, \\ 1/2, & \ell = 0. \end{cases}$$

Starting with a Taylor series expansion

$$U(x,t+\tau) = U(x,t) + \tau\frac{\partial U(x,t)}{\partial t} + \frac{\tau^2}{2}\frac{\partial^2 U(x,t)}{\partial t^2} + O(\tau^3) \tag{3.79}$$

and substituting the integro-differential equation (3.68) and its time derivative, we have

$$U(x,t+\tau) = U(x,t) + \tau\left(f(x,t) + \frac{\tau}{2}\frac{\partial f(x,t)}{\partial t}\right)$$
$$+\frac{\tau^2}{2}a(0)\frac{\partial}{\partial x}\sigma\left(\frac{\partial U(x,t)}{\partial x}\right) \tag{3.80}$$
$$+\tau\int_0^t\left[a(t-s) + \frac{\tau}{2}\frac{\partial a(t-s)}{\partial t}\right]\frac{\partial}{\partial x}\sigma\left(\frac{\partial U(x,t)}{\partial x}\right)ds$$
$$+O(\tau^3).$$

Combining (3.73), (3.74), and (3.80), we have

$$U_i^{j+1} = U_i^j + \tau \left(f_i^j + \frac{\tau}{2} \frac{\partial f(x_i, t_j)}{\partial t} \right)$$
$$+\tau^2 \sum_{\ell=0}^{j} \zeta_\ell a(t_j - t_\ell) \frac{\sigma((U_{i+1}^\ell - U_i^\ell)/h) - \sigma((U_i^\ell - U_{i-1}^\ell)/h)}{h}$$
$$+O(h^2\tau^2 + \tau^3), \tag{3.81}$$

where

$$\zeta_\ell = \begin{cases} 1, & \ell > 0, \\ 1/2, & \ell = 0. \end{cases}$$

Note that using (3.81) one can march forward in time, starting with (3.77). Clearly the method is second order in time. What is maybe surprising is that the only difference between (3.75) and (3.81) is the term $\frac{\tau^2}{2} \frac{\partial f(x_i, t_j)}{\partial t}$.

This method is only conditionally stable and it was shown in [378] following [23] that the condition for stability is

$$1 - 2a(0)\mu \left(\frac{\tau}{h} \right)^2 > 0,$$

where μ is a lower bound for $\sigma'(\xi)$.

We have carried out several numerical experiments using the finite difference scheme (3.78). These experiments demonstrate agreement with the theoretical results (see section 4.4 of Chapter 4).

Now let us construct the semi-discrete analogue of problem (3.68)-(3.70).

Discretization with respect to space variable gives:

$$\frac{du_i}{dt} = \int_0^t a(t-\tau)\left(\sigma\left(u_{\bar{x},i}\right)\right)_x d\tau + f(x_i, t), \qquad (3.82)$$

$$i = 1, 2, \ldots, M-1,$$

$$u_0(t) = u_M(t) = 0, \qquad (3.83)$$

$$u_o(x) = U_{0,i}, \quad i = 0, 1, \ldots, M. \qquad (3.84)$$

So, we obtained a Cauchy problem (3.82)-(3.84) for nonlinear system of ordinary integro-differential equations.

The issue regarding convergence of finite difference as well as semi-discrete schemes, in the best of authors knowledge, is open still today.

3.6 Galerkin's method for Model I

The principal aim of the present section is to construct and study Galerkin's method for approximating the solution of problem (3.1)-(3.3).

Existence and uniqueness of a solution of a first boundary-value problem with nonhomogeneous right-hand side for Model I are examined. As we already said a feature of these equations is the existence, in the coefficients of the highest-order derivatives, of a nonlinear term depending on an integral with respect to time.

This type of equation is a natural generalization, on the one hand, of equations describing applied problems of mathematical physics and, on the other hand, of nonlinear parabolic equations considered by Lions and Vishik in [327], [461] and in many other works as well.

We prove the existence theorem by using Galerkin's method and a compactness procedure developed by M. I. Vishik in the work cited above. The presence of a time integral in the nonlinear terms means that in certain limit procedures, the reasoning must be more delicate than usual in connection with the compactness method.

3.6.1 Preliminary remarks and lemmas

The statements proved in this section for one-dimensional space model are true for multi-dimensional space case and for wider class of nonlinearity [137], [139], [140], [141], [145], [219]. So, preliminary remarks and lemmas are given in a general multidimensional form.

Let Ω be a region in the n-dimensional Euclidean space R^n of the variables $x = (x_1, ..., x_n)$, and let $W_p^k(\Omega)$ denote Sobolev space, i.e., the set of all functions $U(x)$ which, together with their generalized derivatives $D^\alpha U$ of order $|\alpha| \le k$, are in $L_p(\Omega)$, where

$$D^\alpha U = \frac{\partial^\alpha U}{\partial_1^\alpha x_1 \cdots \partial_n^\alpha x_n}, \quad \alpha = (\alpha_1, ..., \alpha_n), \quad |\alpha| = \alpha_1 + \cdots + \alpha_n.$$

When $1 \le p < \infty$, $W_p^k(\Omega)$ becomes Banach space when it is endowed with the norm

$$\|U\|_{W_p^k(\Omega)} = \left[\int_\Omega \left(\sum_{|\alpha| \le k} \left| \frac{\partial^\alpha U}{\partial_1^\alpha x_1 \cdots \partial_n^\alpha x_n} \right|^2 \right)^{p/2} \right]^{1/p}.$$

Let $\mathfrak{D}(\Omega)$ be the set of infinitely differentiable functions with compact support in Ω. As usual $\overset{\circ}{W}{}_p^k(\Omega)$ denotes the closure of $\mathfrak{D}(\Omega)$ in $W_p^k(\Omega)$ with respect to the norm $\|\cdot\|_{W_p^k(\Omega)}$, and $W_q^{-k}(\Omega)$, $1/p + 1/q = 1$, is the space conjugate to $\overset{\circ}{W}{}_p^k(\Omega)$.

We use the notation $L_p\left(0,T;\overset{\circ}{W}{}_p^k(\Omega)\right)$ for the Banach space of functions $V(x,t)$, $x\in\Omega$, $0<t<T$, with the finite norm

$$\|V\|_{L_p\left(0,T;\overset{\circ}{W}{}_p^k(\Omega)\right)}=\left(\int_0^T\|V\|_{\overset{\circ}{W}{}_p^k(\Omega)}\,dt\right)^{1/p}.$$

The following statement takes place (see, for example, [327]).

Lemma 3.1 *Let $\xi\to P(\xi)$ be a continuous mapping of R^n into itself such that, for suitable $\rho>0$, we have $(P(\xi),\xi)\geq 0$ for all ξ from the sphere $|\xi|=\rho$, where, for $\xi=(\xi_1,...,\xi_n)$ and $\eta=(\eta_1,...,\eta_n)$, we put $(\xi,\eta)=\sum\limits_{i=1}^{n}\xi_i\eta_i$ and $|\xi|=(\xi,\xi)^{1/2}$. Then there is ξ, $|\xi|<\rho$, for which $P(\xi)=0$.*

Remark 3.1 *Lemma 3.1 is a variant of Brouwer's fixed-point theorem. We use the lemma in the proof of the existence of solutions of a system of nonlinear algebraic equations obtained by Galerkin's method.*

The following statement is true (see, for example, [461]).

Lemma 3.2 *There is a "basis" of functions $w_1,..,w_m...$, sufficiently smooth in $\overline{Q}_T$, such that the functions Bw_j form a "basis" in $L_p\left(0,T;\overset{\circ}{W}{}_p^k(\Omega)\right)$, where $Q_T=\Omega\times(0,\,T)$,*

$$B\omega=-\frac{\partial}{\partial t}\left[(T-t)\frac{\partial\omega}{\partial t}\right]-\psi\Delta\omega+\lambda\omega,$$

$$\lambda>0,\quad \psi\in\mathfrak{D}(\overline{\Omega}),\quad \psi(x)>0,\quad x\in\Omega,$$
$$\psi=0,\quad \frac{\partial\psi}{\partial n}=0,\quad x\in\partial\Omega.$$

To prove Lemma 3.2 the following two auxiliary problems are considered:

Problem 1. Find a sequence $g_k(t)$, defined on $[0, T]$, such that:

$$-\frac{d}{dt}\left[(T-t)\frac{dg_k}{dt}\right] = \mu_k g_k,$$
$$g_k(0) = 0,$$

$g_k(t)$ is bounded for $t \to T$, the eigenvalues $\mu_k > 0$, and the eigenfunctions $g_k(t)$ are normalized by the condition

$$\int_0^T g_k^2(t)dt = 1.$$

Problem 2. If ξ_m, $m = 1, 2, ...,$ is a "basis" in $\overset{\circ}{W}{}_p^k(\Omega)$ and $\xi_m \in \mathfrak{D}(\Omega)$ find solutions of the problem

$$\psi \Delta V_{km} + (\lambda + \mu_k)V_{km} = \xi_m,$$
$$V_{km} = 0 \quad x \in \partial\Omega.$$

For sufficiently large $\lambda > 0$, problem 2 has a unique solution, smooth in $\overline{\Omega}$.

If we use the diagonal process to enumerate the V_{km} so that they depend on a single index, then we can establish that linear combinations of functions $B(V_{km}g_k)$ can be employed to approximate any function $V(x,t) \in L_p\left(0,\, T;\, \overset{\circ}{W}{}_p^k(\Omega)\right)$, and thus to form a "basis" in $L_p\left(0,\, T;\, \overset{\circ}{W}{}_p^k(\Omega)\right)$.

The following statement takes place [327].

Lemma 3.3 *If Q_T is a bounded domain in $R_x^n \times R_t$ and g_μ and g are functions from $L_q(Q_T)$, $1 < q < \infty$ such that $\|g_\mu\|_{L_q(Q_T)} \le C$ and $g_\mu \to g$ almost everywhere in Q_T, then $g_\mu \to g$ weakly in $L_q(Q_T)$.*

3.6.2 Convergence of Galerkin's method and existence theorem

Let us consider problem (3.1)-(3.3).

Remark 3.2 *For proving that problem (3.1)-(3.3) has a solution, we use the compactness method [327], [461]. Since the equation has the nonlinear term*

$$\frac{\partial}{\partial x}\left\{\left[1+\int_0^t\left(\frac{\partial U}{\partial x}\right)^2 d\tau\right]\frac{\partial U}{\partial x}\right\}$$

we must obtain a-priori estimates of the leading derivatives.

We are looking for a solution $U(x,t)$ of problem (3.1)-(3.3) in $L_4\left(0,\, T;\, \overset{\circ}{W}{}^1_4(\Omega)\right)$ such that $\frac{\partial U}{\partial t} \in L_2(Q_T)$.

Such a solution satisfies the relation

$$\int_0^T \left\langle \frac{\partial U}{\partial t},\, V\right\rangle dt + \int_0^T \langle A(U),\, V\rangle\, dt = \int_0^T \langle f,\, V\rangle\, dt, \quad (3.85)$$

where $V(x,t)$ is an arbitrary function from $L_4\left(0,\, T;\, \overset{\circ}{W}{}^1_4(\Omega)\right)$,

$$A(U) = -\frac{\partial}{\partial x}\left\{\left[1+\int_0^t\left(\frac{\partial U}{\partial x}\right)^2 d\tau\right]\frac{\partial U}{\partial x}\right\} \quad (3.86)$$

and

$$\langle p(x), q(x)\rangle = \int_\Omega p(x)q(x)dx. \quad (3.87)$$

Remark 3.3 *Since the Bw_j form a basis in $L_4\left(0,\, T;\, \overset{\circ}{W}{}^1_4(\Omega)\right)$, it is sufficient to prove that (3.85) holds for a test function of the form $V_j = Bw_j$.*

Theorem 3.5 *If*

$$f,\ \frac{\partial f}{\partial t} \in L_2(Q_T), \quad f(x,\,0) = 0, \tag{3.88}$$

and

$$U_0 = 0, \tag{3.89}$$

then there exists a function $U(x,t)$ with the properties:

$$U \in L_4\left(0,\,T;\,\overset{\circ}{W}{}_4^1(\Omega)\right), \tag{3.90}$$

$$\frac{\partial U}{\partial t} \in L_2(Q_T), \tag{3.91}$$

$$\sqrt{T-t}\sqrt{1+\int_0^t\left(\frac{\partial U}{\partial x}\right)^2 d\tau}\,\frac{\partial^2 U}{\partial t\partial x} \in L_2(Q_T), \tag{3.92}$$

$$\sqrt{\psi}\sqrt{1+\int_0^t\left(\frac{\partial U}{\partial x}\right)^2 d\tau}\,\frac{\partial^2 U}{\partial x^2} \in L_2(Q_T), \tag{3.93}$$

and satisfying (3.1)-(3.3).

Remark 3.4 *Here, in order to simplify the exposition, we are considering only the homogeneous boundary and initial-value problem; however this restriction is in no way essential [327], [461].*

Proof of Theorem 3.5. For proving Theorem 3.5 we are applying the Galerkin's method and looking for a function $u_m(x,t) = \sum_{k=1}^{m} c_k\omega_k(x,t)$ for which

$$\int_0^T \left\langle \frac{\partial u_m}{\partial t} + A(u_m),\, B\omega_j \right\rangle dt = \int_0^T \langle f,\, B\omega_j\rangle\, dt, \tag{3.94}$$

$$1 \le j \le m.$$

Before proving that the system of nonlinear algebraic equations (3.94) has a solution, we establish the necessary a-priori estimates.

The following statement is true.

Lemma 3.4 *Under the conditions of Theorem 3.5 the following estimation takes place*

$$\int_0^T \left\langle \frac{\partial u_m}{\partial t} + A(u_m),\, Bu_m \right\rangle dt \geq C \left\| \frac{\partial u_m}{\partial x} \right\|^2_{L_2(Q_T)}$$

$$+ \left\| \sqrt{T-t} \sqrt{1 + \int_0^t \left(\frac{\partial u_m}{\partial x} \right)^2 d\tau} \frac{\partial^2 u_m}{\partial x\, \partial t} \right\|^2_{L_2(Q_T)}$$

$$+C \left\| \sqrt{\psi} \sqrt{1 + \int_0^t \left(\frac{\partial u_m}{\partial x} \right)^2 d\tau} \frac{\partial^2 u_m}{\partial x^2} \right\|^2_{L_2(Q_T)} \tag{3.95}$$

$$+C \left\| \sqrt{\psi} \int_0^T \frac{\partial}{\partial x} \left(\frac{\partial u_m}{\partial x} \right)^2 dt \right\|^2_{L_2(Q_T)}$$

$$+C \left\| \frac{\partial u_m}{\partial t} \right\|^2_{L_2(Q_T)} + \frac{1}{4} \left\| \frac{\partial u_m}{\partial x} \right\|^4_{L_4(Q_T)}.$$

Here and below we write C for various constants not depending on u_m.

Proof. First of all let us introduce the notations:

$$\int_0^T \left\langle \frac{\partial u_m}{\partial t} + A(u_m),\, Bu_m \right\rangle dt = \sum_{k=1}^{6} J_k,$$

where

$$J_1 = \int_0^T \left\langle \frac{\partial u_m}{\partial t}, -\frac{\partial}{\partial t}\left[(T-t)\frac{\partial u_m}{\partial t}\right]\right\rangle dt$$
$$\geq \frac{1}{2}\left\|\frac{\partial u_m}{\partial t}\right\|^2_{L_2(Q_T)}, \tag{3.96}$$

$$J_2 = \int_0^T \left\langle \frac{\partial u_m}{\partial t}, \psi\frac{\partial^2 u_m}{\partial x^2}\right\rangle dt$$
$$= \int_0^T \left\langle \psi\frac{\partial^2 u_m}{\partial x\,\partial t}, \frac{\partial u_m}{\partial x}\right\rangle dt + \int_0^T \left\langle \frac{\partial u_m}{\partial t}\frac{\partial \psi}{\partial x}, \frac{\partial u_m}{\partial x}\right\rangle dt \tag{3.97}$$
$$\geq \frac{1}{4}\left\|\frac{\partial u_m}{\partial t}\right\|^2_{L_2(Q_T)} - C\left\|\frac{\partial u_m}{\partial x}\right\|^2_{L_2(Q_T)},$$

$$J_3 = \lambda\int_0^T \left\langle \frac{\partial u_m}{\partial t}, u_m\right\rangle dt = \frac{\lambda}{2}\int_{Q_T}\frac{\partial u_m^2}{\partial t}dxdt \geq 0, \tag{3.98}$$

$$J_4 = \int_0^T \left\langle -\frac{\partial}{\partial x}\left\{\left[1+\int_0^t \left(\frac{\partial u_m}{\partial x}\right)^2 d\tau\right]\frac{\partial u_m}{\partial x}\right\}\right.$$
$$\left.-\frac{\partial}{\partial t}\left[(T-t)\frac{\partial u_m}{\partial t}\right]\right\rangle dt = \int_0^T \left\langle \psi\frac{\partial^2 u_m}{\partial x\,\partial t}, \frac{\partial u_m}{\partial x}\right\rangle dt$$
$$+\int_0^T \left\langle \frac{\partial u_m}{\partial t}\frac{\partial \psi}{\partial x}, \frac{\partial u_m}{\partial x}\right\rangle dt,$$

$$J_4 = \int_0^T \left\langle (T-t)\left[1+\int_0^t \left(\frac{\partial u_m}{\partial x}\right)^2 d\tau\right], \left(\frac{\partial^2 u_m}{\partial x\,\partial t}\right)^2 \right\rangle dt$$

$$+\int_{Q_T} \frac{T-t}{4}\frac{\partial}{\partial t}\left(\frac{\partial u_m}{\partial x}\right)^2 dxdt$$

$$= \left\| \sqrt{T-t}\sqrt{1+\int_0^t \left(\frac{\partial u_m}{\partial x}\right)^2 d\tau}\frac{\partial^2 u_m}{\partial x\,\partial t}\right\|^2_{L_2(Q_T)} + \frac{1}{4}\left\|\frac{\partial u_m}{\partial x}\right\|^4_{L_2(Q_T)},$$

$$J_5 = \int_0^T \left\langle -\frac{\partial}{\partial x}\left\{\left[1+\int_0^t \left(\frac{\partial u_m}{\partial x}\right)^2 d\tau\right]\frac{\partial u_m}{\partial x}\right\}, -\psi\frac{\partial^2 u_m}{\partial x^2}\right\rangle dt$$

$$= \frac{1}{2}\int_{Q_T}\psi\int_0^t \frac{\partial}{\partial x}\left(\frac{\partial u_m}{\partial x}\right)^2 d\tau\left(\frac{\partial u_m}{\partial x}\right)^2 dxdt$$

$$+\int_{Q_T}\psi\left[1+\int_0^t \left(\frac{\partial u_m}{\partial x}\right)^2 d\tau\right]\left(\frac{\partial^2 u_m}{\partial x^2}\right)^2 dxdt. \tag{3.99}$$

Using the notation

$$\varphi(x,t) = \int_0^t \frac{\partial}{\partial x}\left(\frac{\partial u_m}{\partial x}\right)^2 d\tau,$$

we have

$$J_5 = \int_{Q_T}\psi\int_0^t \frac{\partial}{\partial x}\left(\frac{\partial u_m}{\partial x}\right)^2 d\tau\left(\frac{\partial u_m}{\partial x}\right)^2 dxdt$$

$$+\frac{1}{4}\int_{Q_T}\psi\frac{\partial\varphi^2}{\partial t}dxdt$$

$$J_5 = \left\| \sqrt{\psi}\sqrt{1+\int_0^t \left(\frac{\partial u_m}{\partial x}\right)^2 d\tau}\,\frac{\partial^2 u_m}{\partial x^2} \right\|^2_{L_2(Q_T)}$$

$$+\frac{1}{4}\left\| \sqrt{\psi}\int_0^T \frac{\partial}{\partial x}\left(\frac{\partial^2 u_m}{\partial x^2}\right) dt \right\|^2_{L_2(Q_T)}.$$

$$J_6 = -\int_0^T \left\langle \frac{\partial}{\partial x}\left\{\left[1+\int_0^t \left(\frac{\partial u_m}{\partial x}\right)^2 d\tau\right]\frac{\partial u_m}{\partial x}\right\}, \lambda u_m \right\rangle dt$$

$$= \int_{Q_T}\left[1+\int_0^t \left(\frac{\partial u_m}{\partial x}\right)^2 d\tau\right]\left(\frac{\partial u_m}{\partial x}\right)^2 dxdt$$

$$= \lambda\int_{Q_T}\left(\frac{\partial u_m}{\partial x}\right)^2 dxdt + \frac{\lambda}{2}\int_{Q_T}\frac{\partial \varphi_1}{\partial t} dxdt, \tag{3.100}$$

where

$$\varphi_1(x,t) = \int_0^t \left(\frac{\partial u_m}{\partial x}\right)^2 d\tau.$$

Furthermore

$$J_6 = \lambda\int_{Q_T}\left(\frac{\partial u_m}{\partial x}\right)^2 dxdt$$

$$+\frac{\lambda}{2}\int_{\Omega}\left[\int_0^T \left(\frac{\partial u_m}{\partial x}\right)^2 dt\right]^2 dx \tag{3.101}$$

$$= \lambda\left\|\frac{\partial u_m}{\partial x}\right\|^2_{L_2(Q_T)} + \frac{\lambda}{2}\left\|\int_0^T \left(\frac{\partial u_m}{\partial x}\right)^2 dt\right\|^2_{L_2(Q_T)}.$$

Combining the estimates of J_k, $k = 1, ..., 6$, we obtain the assertion of Lemma 3.4 for $\lambda > C$. The right side of relation (3.94) satisfies the inequality

$$\left|\int_0^T \langle f,\, Bu_m\rangle\, dt\right| = \left|\int_0^T \left\langle f,\, -\frac{\partial}{\partial t}\left[(T-t)\frac{\partial u_m}{\partial t}\right]\right.\right.$$

$$\left.\left. -\psi\frac{\partial^2 u_m}{\partial x^2} + \lambda u_m\right\rangle dt\right|$$

$$\leq \left|\int_0^T \left\langle \frac{\partial f}{\partial t},\, (T-t)\frac{\partial u_m}{\partial t}\right\rangle dt\right|$$

$$+\left|\int_0^T \left\langle f,\, -\psi\frac{\partial^2 u_m}{\partial x^2}\right\rangle dt\right| + \left|\int_0^T \langle f,\, \lambda u_m\rangle\, dt\right| \tag{3.102}$$

$$+C\left\|\frac{\partial u_m}{\partial t}\right\|^2_{L_2(Q_T)} + C\left\|\sqrt{\psi}\frac{\partial^2 u_m}{\partial x^2}\right\|^2_{L_2(Q_T)}$$

$$+C\left\|\frac{\partial u_m}{\partial x}\right\|^2_{L_2(Q_T)} + C(f).$$

Here the coefficients C can vary, and $C(f)$ is a constant depending on $f(x,t)$. Using (3.102) and the inequalities in Lemma 3.4, we conclude that

$$\int_0^T \left\langle \frac{\partial u_m}{\partial t} + A(u_m) - f,\, Bu_m\right\rangle dt \geq 0 \tag{3.103}$$

when the following sum is sufficiently large

$$\left\|\frac{\partial u_m}{\partial t}\right\|^2_{L_2(Q_T)} + \left\|\frac{\partial u_m}{\partial x}\right\|^4_{L_4(Q_T)}.$$

Remark 3.5 *For proving (3.103) we use Young's inequality* $ab \le \varepsilon a^p + c(\varepsilon) b^q$, $\frac{1}{p} + \frac{1}{q} = 1$. *By virtue of Lemma 3.1, there exists a function* $u_m(x,t)$ *satisfying (3.94); hence,*

$$\left\| \sqrt{T-t}\sqrt{1+\int_0^t \left(\frac{\partial u_m}{\partial x}\right)^2 d\tau}\,\frac{\partial^2 u_m}{\partial x\,\partial t}\right\|^2_{L_2(Q_T)} + \left\|\frac{\partial u_m}{\partial t}\right\|^2_{L_2(Q_T)} + \left\|\sqrt{\psi}\sqrt{1+\int_0^t \left(\frac{\partial u_m}{\partial x}\right)^2 d\tau}\,\frac{\partial^2 u_m}{\partial x^2}\right\|^2_{L_2(Q_T)} + \left\|\frac{\partial u_m}{\partial x}\right\|^4_{L_4(Q_T)} + \left\|\sqrt{\psi}\int_0^T \frac{\partial}{\partial x}\left(\frac{\partial u_m}{\partial x}\right)^2 dt\right\|^2_{L_2(Q_T)} + \left\|\int_0^T \left(\frac{\partial u_m}{\partial x}\right)^2 dt\right\|^2_{L_2(Q_T)} < C. \tag{3.104}$$

To estimate the nonlinear term, we use the following result.

Lemma 3.5 *If* $V \in L_4\left(0,T;\overset{\circ}{W}{}^1_4(\Omega)\right)$, *then*

$$\left\| \left[1+\int_0^t \left(\frac{\partial V}{\partial x}\right)^2 d\tau\right] \frac{\partial V}{\partial x}\right\|_{L_{4/3}(Q_T)} < \infty. \tag{3.105}$$

Proof. It is sufficient to estimate

$$\int_0^t \left(\frac{\partial V}{\partial x}\right)^2 d\tau \frac{\partial V}{\partial x}.$$

First,

$$\int_{Q_T}\left[\int_0^t\left(\frac{\partial V}{\partial x}\right)^2 d\tau\right]^{\frac{4}{3}}\left|\frac{\partial V}{\partial x}\right|^{\frac{4}{3}} dxdt$$

$$\leq C\int_{Q_T}\left[\int_0^T\left(\frac{\partial V}{\partial x}\right)^4 d\tau\right]^{\frac{2}{3}}\left|\frac{\partial V}{\partial x}\right|^{\frac{4}{3}} dxdt$$

$$\leq C\left[\int_{Q_T}\int_0^T\left(\frac{\partial V}{\partial x}\right)^4 d\tau\right]^{\frac{2}{3}}\left|\frac{\partial V}{\partial x}\right|^{\frac{4}{3}} dxdt$$

$$\times\left[\int_{Q_T}\left(\frac{\partial V}{\partial x}\right)^4\right]^{\frac{1}{3}}\left|\frac{\partial V}{\partial x}\right|^{\frac{4}{3}} dxdt$$

$$= CT^{\frac{2}{3}}\int_{Q_T}\left(\frac{\partial V}{\partial x}\right)^4 dxdt < \infty.$$

Therefore, the operation $V \to A(V)$ maps $V \in L_4\left(0, T; \overset{\circ}{W}{}^1_4(\Omega)\right)$ into $A(V) \in L_{4/3}\left(0, T; \overset{\circ}{W}{}^{-1}_{4/3}(\Omega)\right)$. Lemma 3.5 implies that

$$\left\|\left[1+\int_0^t\left(\frac{\partial u_m}{\partial x}\right)^2 d\tau\right]\frac{\partial u_m}{\partial x}\right\|_{L_{4/3}(Q_T)} \leq C, \qquad (3.106)$$
$$m = 1,\ 2,\ \ldots$$

Now we are going to estimate u_m and study its convergence. From (3.104) and (3.106) it follows that, for $m \to \infty$:

$$u_m \quad \text{are bounded in} \quad L_4\left(0, T; \overset{\circ}{W}{}^1_4(\Omega)\right), \qquad (3.107)$$

$$\frac{\partial u_m}{\partial t} \quad \text{are bounded in} \quad L_2(Q_T), \qquad (3.108)$$

$$\sqrt{T-t}\sqrt{1+\int_0^t\left(\frac{\partial u_m}{\partial x}\right)^2 d\tau}\,\frac{\partial^2 u_m}{\partial x\,\partial t} \quad \text{are bounded in} \quad L_2(Q_T), \tag{3.109}$$

$$\sqrt{\psi}\sqrt{1+\int_0^t\left(\frac{\partial u_m}{\partial x}\right)^2 d\tau}\,\frac{\partial^2 u_m}{\partial x^2} \quad \text{are bounded in} \quad L_2(Q_T), \tag{3.110}$$

$$\left[1+\int_0^t\left(\frac{\partial u_m}{\partial x}\right)^2 d\tau\right]\frac{\partial u_m}{\partial x} \quad \text{are bounded in} \quad L_{4/3}(Q_T), \tag{3.111}$$

$$\sqrt{\psi}\int_0^T \frac{\partial}{\partial x}\left(\frac{\partial u_m}{\partial x}\right)^2 dt \quad \text{are bounded in} \quad L_2(Q_T), \tag{3.112}$$

$$\int_0^T \left(\frac{\partial u_m}{\partial x}\right)^2 dt \quad \text{are bounded in} \quad L_2(Q_T). \tag{3.113}$$

The following relations are consequences of (3.107)-(3.112) [327]:

$$u_m \to U \quad \text{weakly in} \quad L_4\left(0,T;\overset{\circ}{W}{}^1_4(\Omega)\right), \tag{3.114}$$

$$\frac{\partial u_m}{\partial t} \to \frac{\partial U}{\partial t} \quad \text{weakly in} \quad L_2\left(Q_T\right), \tag{3.115}$$

$$\frac{\partial u_m}{\partial x} \to \frac{\partial U}{\partial x} \quad \text{almost everywhere in} \quad Q_T, \tag{3.116}$$

$$\sqrt{T-t}\sqrt{1+\int_0^t\left(\frac{\partial u_m}{\partial x}\right)^2 d\tau}\,\frac{\partial^2 u_m}{\partial x\,\partial t} \to \chi \quad \text{weakly in} \quad L_2\left(Q_T\right), \tag{3.117}$$

$$\sqrt{\psi}\sqrt{1+\int_0^t\left(\frac{\partial u_m}{\partial x}\right)^2 d\tau}\,\frac{\partial^2 u_m}{\partial x^2} \to \eta \quad \text{weakly in} \quad L_2\left(Q_T\right), \tag{3.118}$$

$$\left[1+\int_0^t \left(\frac{\partial u_m}{\partial x}\right)^2 d\tau\right] \frac{\partial u_m}{\partial x} \to \xi \quad \text{weakly in} \quad L_{4/3}\left(Q_T\right), \tag{3.119}$$

$$\int_0^T \left(\frac{\partial u_m}{\partial x}\right)^2 dt \to \zeta \quad \text{almost everywhere in} \quad \Omega. \tag{3.120}$$

To be able to take the limit in the nonlinear term in (3.94), we should prove that

$$\int_0^t \left(\frac{\partial u_m}{\partial x}\right)^2 d\tau \to \int_0^t \left(\frac{\partial U}{\partial x}\right)^2 d\tau \quad \text{almost everywhere in} \quad Q_T. \tag{3.121}$$

At first we prove that (3.116) can be strengthened, i.e., we show that

$$\frac{\partial u_m}{\partial x} \to \frac{\partial U}{\partial x} \quad \text{weakly in} \quad L_2(Q_T). \tag{3.122}$$

In fact, by virtue of de la Vallee Poussin's theorem [374] and (3.107), the sequence of integrals

$$\int_{Q_T} \left(\frac{\partial u_m}{\partial x}\right)^2 dx dt$$

is absolutely equicontinuous; hence, by virtue of Vitali's theorem [374] we have

$$\lim_{m\to\infty} \int_{Q_T} \left(\frac{\partial u_m}{\partial x}\right)^2 dx dt = \int_{Q_T} \left(\frac{\partial U}{\partial x}\right)^2 dx dt$$

or, equivalently,

$$\left\|\frac{\partial u_m}{\partial x}\right\|_{L_2(Q_T)} \to \left\|\frac{\partial U}{\partial x}\right\|_{L_2(Q_T)}. \tag{3.123}$$

Moreover, application of (3.116) and Lemma 3.3 establishes that

$$\frac{\partial u_m}{\partial x} \to \frac{\partial U}{\partial x} \quad \text{weakly in} \quad L_2(Q_T).$$

Thus from Riesz's theorem it follows that (3.122) holds.

The definition of mean convergence implies that

$$\lim_{m\to\infty} \int_{Q_T} \left(\frac{\partial u_m}{\partial x} - \frac{\partial U}{\partial x} \right)^2 dxdt = 0. \tag{3.124}$$

Let us introduce the following notation

$$g_m = \int_0^T \left(\frac{\partial u_m}{\partial x} - \frac{\partial U}{\partial x} \right)^2 dt.$$

Relations (3.120) and (3.124) imply that the sequence $g_m(x)$ converges in $L_1(\Omega)$; there is therefore a subsequence of functions $g_m(x)$ converging to zero for almost all $x \in \Omega$, i.e.,

$$\int_0^T \left(\frac{\partial u_m}{\partial x} - \frac{\partial U}{\partial x} \right)^2 dt \to 0. \tag{3.125}$$

or, equivalently,

$$\frac{\partial u_m}{\partial x} \to \frac{\partial U}{\partial x} \quad \text{in} \quad L_2(0,T)$$

for almost all $x \in \Omega$; hence,

$$\int_0^T \left(\frac{\partial u_m}{\partial x} \right)^2 dt \to \int_0^T \left(\frac{\partial U}{\partial x} \right)^2 dt. \tag{3.126}$$

From this follows relation (3.121), because we can take the limit under the integral sign on each subset $[0,t] \subset [0,T]$.

We conclude from (3.119), (3.121), and Lemma 3.3 that:

$$\xi = \left[1 + \int_0^t \left(\frac{\partial U}{\partial x}\right)^2 d\tau\right] \frac{\partial U}{\partial x},$$

$$\chi = \sqrt{T-t}\sqrt{1 + \int_0^t \left(\frac{\partial U}{\partial x}\right)^2 d\tau}\,\frac{\partial^2 U}{\partial x\, \partial t},$$

$$\eta = \sqrt{\psi}\sqrt{1 + \int_0^t \left(\frac{\partial U}{\partial x}\right)^2 d\tau}\,\frac{\partial^2 U}{\partial x^2}.$$

Hence,

$$A(u_m) \to A(U) \quad in \quad L_{4/3}\left(0, T; \overset{\circ}{W}{}^{-1}_{4/3}(\Omega)\right).$$

Equalities (3.94) imply that

$$\int_0^T \left\langle \frac{\partial U}{\partial t} + A(U),\, Bw_j \right\rangle dt = \int_0^T \langle f,\, B\omega_j\rangle\, dt, \quad j = 1, 2, \ldots$$

and the existence result, Theorem 3.5, has been proved.

3.6.3 Uniqueness of solution

Now we are going to prove uniqueness of solution of problem (3.1)-(3.3). Let U_1 and U_2 satisfy (3.1)-(3.3); if $W = U_1 - U_2$, then

$$\frac{\partial W}{\partial t} + A(U_1) - A(U_2) = 0. \tag{3.127}$$

Multiplying (3.127) by W and applying elementary transformation, we obtain the relation

$$\int_0^t \int_\Omega \left\{ W\frac{\partial W}{\partial t} - \left\{ \frac{\partial}{\partial x}\left[\left(1 + \int_0^{t'} \left(\frac{\partial U_1}{\partial x}\right)^2 d\tau\right)\frac{\partial U_1}{\partial x}\right]\right.\right.$$

$$-\frac{\partial}{\partial x}\left[\left(1+\int_0^{t'}\left(\frac{\partial U_2}{\partial x}\right)^2 d\tau\right)\frac{\partial U_2}{\partial x}\right]\Bigg\} W\Bigg\} dxdt'$$

$$=\frac{1}{2}\|W(t)\|^2_{L_2(\Omega)}+\int_0^t\left\|\frac{\partial W}{\partial x}\right\|^2_{L_2(\Omega)}dt$$

$$+\int_\Omega\int_0^t\left[\int_0^{t'}\left(\frac{\partial U_1}{\partial x}\right)^2 d\tau\left(\frac{\partial U_1}{\partial x}\right)^2\right.$$
$$\left.+\int_0^{t'}\left(\frac{\partial U_2}{\partial x}\right)^2 d\tau\left(\frac{\partial U_2}{\partial x}\right)^2\right]dxdt$$

$$-\int_\Omega\int_0^t\int_0^{t'}\left[\left(\frac{\partial U_1}{\partial x}\right)^2+\left(\frac{\partial U_2}{\partial x}\right)^2\right]d\tau\frac{\partial U_1}{\partial x}\frac{\partial U_2}{\partial x}dxdt'=0.$$

Simplification yields

$$\|W(t)\|^2_{L_2(\Omega)}+\int_0^t\int_\Omega\frac{\partial\varphi^2(x,t')}{\partial t'}dxdt'\leq 0,$$

where

$$\varphi(x,t)=\int_0^t\left[\left(\frac{\partial U_1}{\partial x}\right)^2-\left(\frac{\partial U_2}{\partial x}\right)^2\right]d\tau;$$

hence,

$$\|W(t)\|^2_{L_2(\Omega)}+\|\varphi(t)\|^2_{L_2(\Omega)}\leq 0$$

and so $W(x,\,t)\equiv 0$. Therefore, $U_1(x,\,t)\equiv U_2(x,\,t)$ and uniqueness of solution of problem (3.1)-(3.3) has been proved.

3.7 Galerkin's method for Model II

The principal aim of this section is to study Galerkin's method for problem (3.34)-(3.36).

3.7.1 Variational formulation

One of the ingredients of the finite element method is a variational formulation of the problem. To provide this variational formulation, let us denote by H the linear space of functions u satisfying the boundary conditions in (3.2) and

$$||u(\cdot,t)||_1 < \infty,$$

where

$$||u(\cdot,t)||_r = \left\{ \int_0^1 \left[|u(x,t)|^2 + \sum_{j=1}^{r} \left| \frac{\partial^j u(x,t)}{\partial x^j} \right|^2 \right] dx \right\}^{1/2}. \tag{3.128}$$

Whenever $r = 0$ we will omit the subscript.

The variational formulation of problem (3.34)-(3.36) can be stated as follows:

Find a function $U(x,t) \in H$ for which

$$\begin{aligned} &\left\langle V, \frac{\partial U}{\partial t} \right\rangle + \left\langle (1+S(t))\frac{\partial U}{\partial x}, \frac{\partial V}{\partial x} \right\rangle \\ &= \langle f, V \rangle, \qquad \forall V \in H, \end{aligned} \tag{3.129}$$

and

$$\langle V, U(x,0) \rangle = \langle V, u_0(x) \rangle, \qquad \forall V \in H, \tag{3.130}$$

where

$$S(t) = \int_0^t \int_0^1 \left(\frac{\partial U}{\partial x} \right)^2 dx d\tau$$

and the inner product is given by (3.87).

To approximate the solution of (3.129), (3.130) we require that U lies in a finite-dimensional subspace H^h of H for each t (see, e.g., [375]). The following property concerning approximability in H^h can be readily verified for finite element spaces, see [439].

Approximation Property

There is an integer $r \geq 2$ and positive numbers C_0, C_1 independent of h such that for any $V \in H$ there exists a point $v^h \in H^h$ satisfying

$$\begin{gathered} ||V - v^h||_\ell \leq C_\ell h^{j-\ell} ||V||_j \\ \text{for} \quad 0 \leq \ell \leq 1, \quad \ell < j \leq r. \end{gathered} \tag{3.131}$$

Approximate Problem

The approximation $u^h \in H^h$ to U is defined by the following variational analogue of (3.129), (3.130):
Find a $u^h \in H^h$ such that

$$\begin{aligned} &\left\langle v^h, \frac{\partial u^h}{\partial t} \right\rangle + \left\langle (1 + S_h(t)) \frac{\partial u^h}{\partial x}, \frac{\partial v^h}{\partial x} \right\rangle \\ &= \left\langle f, v^h \right\rangle, \qquad \forall v^h \in H^h, \end{aligned} \tag{3.132}$$

and

$$\left\langle v^h, u^h(x,0) \right\rangle = \left\langle v^h, U_0(x) \right\rangle, \qquad \forall v^h \in H^h, \tag{3.133}$$

where

$$S_h(t) = \int_0^t \int_0^1 \left(\frac{\partial u^h}{\partial x} \right)^2 dx d\tau.$$

Once a basis has been selected for H^h, (3.132), (3.133) are equivalent to a set of N integro-differential equations, where N is dimension of H^h. The solution to this system will be discussed in section 4.3 of Chapter 4.

3.7.2 Error estimates

In this section we shall estimate the error in the finite element approximation using the norm

$$|||E|||_r = \left\{ \int_0^T \int_0^1 \sum_{i=0}^{r} \left| \frac{\partial^i E(x,t)}{\partial x^i} \right|^2 dx dt \right\}^{1/2} . \qquad (3.134)$$

Everywhere in the case $r = 0$ we will omit the subscript. Below C and c_i denote various constants independent of h.

Theorem 3.6 *The error in the finite element approximation u^h generated by (3.132), (3.133) satisfies the relation*

$$\begin{aligned} |||U - u^h|||_1 \leq & \; h^{j-1} C \left\{ \frac{1}{2} h^2 ||U_0||^2 + C h^2 |||U_t|||^2 \right. \\ + & \; C^2 \left[1 + h^{2(j-1)} |||U|||^2 \right] |||U|||^2 \\ + & \; \left. C^3 \left(h^{j-1} |||U|||^2 + []U[] \right)^2 \right\}^{1/2} , \end{aligned}$$

where

$$[]U[] = \int_0^T \int_0^1 |U| dx d\tau .$$

Proof. Subtracting (3.132) from (3.129) with v^h instead of V, we obtain

$$\left\langle v^h, \frac{\partial u^h}{\partial t} \right\rangle + \left\langle (1 + S_h(t)) \frac{\partial u^h}{\partial x}, \frac{\partial v^h}{\partial x} \right\rangle = \left\langle v^h, \frac{\partial U}{\partial t} \right\rangle$$

$$+ \left\langle (1 + S(t)) \frac{\partial U}{\partial x}, \frac{\partial v^h}{\partial x} \right\rangle , \qquad \forall v^h \in H^h .$$

Let $\tilde{u}^h$ be any function in H^h, then

$$\begin{aligned}
&\left\langle v^h, \frac{\partial\left(u^h - \tilde{u}^h\right)}{\partial t}\right\rangle \\
&+\left\langle\left[(1+S_h(t))\frac{\partial u^h}{\partial x} - (1+\tilde{S}_h(t))\frac{\partial \tilde{u}^h}{\partial x}\right], \frac{\partial v^h}{\partial x}\right\rangle \\
&=\left\langle v^h, \frac{\partial\left(U - \tilde{u}^h\right)}{\partial t}\right\rangle \\
&+\left\langle\left[(1+S(t))\frac{\partial U}{\partial x} - (1+\tilde{S}_h(t))\frac{\partial \tilde{u}^h}{\partial x}\right], \frac{\partial v^h}{\partial x}\right\rangle, \\
&\qquad \forall v^h \in H^h,
\end{aligned} \tag{3.135}$$

where

$$\tilde{S}_h(t) = \int_0^t \int_0^1 \left(\frac{\partial \tilde{u}^h}{\partial x}\right)^2 dx d\tau.$$

Let us define the errors as follows:

$$\begin{aligned}
e(x,t) &= u^h(x,t) - \tilde{u}^h(x,t), \\
E(x,t) &= U(x,t) - \tilde{u}^h(x,t).
\end{aligned} \tag{3.136}$$

Since $e \in H^h$, we can let $v^h = e$ and (3.135) becomes

$$\begin{aligned}
&\left\langle e, \frac{\partial e}{\partial t} \right\rangle \\
&+ \left\langle \left[(1 + S_h(t)) \frac{\partial u^h}{\partial x} - (1 + \tilde{S}_h(t)) \frac{\partial \tilde{u}^h}{\partial x} \right], \frac{\partial e}{\partial x} \right\rangle \\
&= \left\langle e, \frac{\partial E}{\partial t} \right\rangle \\
&+ \left\langle \left[(1 + S(t)) \frac{\partial U}{\partial x} - (1 + \tilde{S}_h(t)) \frac{\partial \tilde{u}^h}{\partial x} \right], \frac{\partial e}{\partial x} \right\rangle .
\end{aligned} \tag{3.137}$$

Let us consider the second term on the left of (3.137)

$$\begin{aligned}
&\left\langle \left[(1 + S_h(t)) \frac{\partial u^h}{\partial x} - (1 + \tilde{S}_h(t)) \frac{\partial \tilde{u}^h}{\partial x} \right], \frac{\partial e}{\partial x} \right\rangle \\
&= \left\langle \frac{\partial e}{\partial x}, \frac{\partial e}{\partial x} \right\rangle + \left\langle \left[S_h(t) \frac{\partial u^h}{\partial x} - \tilde{S}_h(t) \frac{\partial \tilde{u}^h}{\partial x} \right], \frac{\partial e}{\partial x} \right\rangle .
\end{aligned} \tag{3.138}$$

Denoting:

$$\omega(x, t) = \frac{\partial u^h(x, t)}{\partial x}, \qquad \eta(x, t) = \frac{\partial \tilde{u}^h(x, t)}{\partial x},$$

the last term of (3.138) can be rewritten as

$$\left\langle \left[S_h(t)\frac{\partial u^h}{\partial x} - \tilde{S}_h(t)\frac{\partial \tilde{u}^h}{\partial x} \right], \frac{\partial e}{\partial x} \right\rangle$$

$$= \left\langle \omega \int_0^t \int_0^1 \omega^2 dx d\tau - \eta \int_0^t \int_0^1 \eta^2 dx d\tau,\ \omega - \eta \right\rangle$$

$$= \frac{1}{2} \left\langle \int_0^t \int_0^1 \omega^2 dx d\tau + \int_0^t \int_0^1 \eta^2 dx d\tau,\ (\omega - \eta)^2 \right\rangle$$

$$+ \frac{1}{2} \left\langle \int_0^t \int_0^1 \omega^2 dx d\tau - \int_0^t \int_0^1 \eta^2 dx d\tau,\ \omega^2 - \eta^2 \right\rangle$$

$$\geq \frac{1}{2} \left\langle \int_0^t \int_0^1 \omega^2 dx d\tau - \int_0^t \int_0^1 \eta^2 dx d\tau,\ \omega^2 - \eta^2 \right\rangle$$

$$= \frac{1}{2} \int_0^1 \int_0^t \int_0^1 \left(\omega^2 - \eta^2\right) dx d\tau \left(\omega^2 - \eta^2\right) dx$$

$$= \frac{1}{2} \int_0^t \int_0^1 \left(\omega^2 - \eta^2\right) dx d\tau \int_0^1 \left(\omega^2 - \eta^2\right) dx = \frac{1}{4}\frac{d\varphi^2(t)}{dt},$$

where

$$\varphi(t) = \int_0^t \int_0^1 \left(\omega^2 - \eta^2\right) dx d\tau.$$

Therefore, the left-hand side of (3.137) can be rewritten as follows

$$\left\langle e, \frac{\partial e}{\partial t} \right\rangle + \left\langle \left[(1 + S_h(t))\frac{\partial u^h}{\partial x} - (1 + \tilde{S}_h(t))\frac{\partial \tilde{u}^h}{\partial x} \right], \frac{\partial e}{\partial x} \right\rangle$$

$$\geq \frac{1}{2}\frac{d}{dt}\|e\|^2 + \left\| \frac{\partial e}{\partial x} \right\|^2 + \frac{1}{4}\frac{d\varphi^2(t)}{dt}.$$

Now consider the second term on the right of (3.137)

$$\left\langle \left[(1+S(t))\frac{\partial U}{\partial x} - (1+\tilde{S}_h(t))\frac{\partial \tilde{u}^h}{\partial x} \right], \frac{\partial e}{\partial x} \right\rangle$$
$$= \left\langle \frac{\partial E}{\partial x}, \frac{\partial e}{\partial x} \right\rangle + \left\langle \left[S(t)\frac{\partial U}{\partial x} - \tilde{S}_h(t)\frac{\partial \tilde{u}^h}{\partial x} \right], \frac{\partial e}{\partial x} \right\rangle . \tag{3.139}$$

Substituting for $\frac{\partial U}{\partial x}$ from (3.136), the last equality gives

$$S(t)\frac{\partial U}{\partial x} - \tilde{S}_h(t)\frac{\partial \tilde{u}^h}{\partial x}$$
$$= \left(\frac{\partial E}{\partial x} + \frac{\partial \tilde{u}^h}{\partial x} \right) \int_0^t \int_0^1 \left(\frac{\partial E}{\partial x} + \frac{\partial \tilde{u}^h}{\partial x} \right)^2 dx d\tau$$
$$- \frac{\partial \tilde{u}^h}{\partial x} \int_0^t \int_0^1 \left(\frac{\partial \tilde{u}^h}{\partial x} \right)^2 dx d\tau$$
$$= \left(\frac{\partial E}{\partial x} + \frac{\partial \tilde{u}^h}{\partial x} \right) \int_0^t \int_0^1 \left[\left(\frac{\partial E}{\partial x} \right)^2 + 2\frac{\partial E}{\partial x}\frac{\partial \tilde{u}^h}{\partial x} \right] dx d\tau$$
$$+ \frac{\partial E}{\partial x} \int_0^t \int_0^1 \left(\frac{\partial \tilde{u}^h}{\partial x} \right)^2 dx d\tau .$$

Taking this into account in the right-hand side of (3.137), we get

$$\left\langle e, \frac{\partial E}{\partial t} \right\rangle + \left\langle \frac{\partial E}{\partial x}, \frac{\partial e}{\partial x} \right\rangle + \left\langle \frac{\partial E}{\partial x}, \frac{\partial e}{\partial x} \right\rangle \int_0^t \int_0^1 \left(\frac{\partial \tilde{u}^h}{\partial x} \right)^2 dx d\tau$$

$$+ \left\langle \frac{\partial E}{\partial x}, \frac{\partial e}{\partial x} \right\rangle \int_0^t \int_0^1 \left(\frac{\partial E}{\partial x} \right)^2 dx d\tau$$

$$+ \left\langle \frac{\partial \tilde{u}^h}{\partial x}, \frac{\partial e}{\partial x} \right\rangle \int_0^t \int_0^1 \left(\frac{\partial E}{\partial x} \right)^2 dx d\tau$$

$$+ \left\langle \frac{\partial E}{\partial x} + \frac{\partial \tilde{u}^h}{\partial x}, \frac{\partial e}{\partial x} \right\rangle \int_0^t \int_0^1 2 \frac{\partial \tilde{u}^h}{\partial x} \frac{\partial E}{\partial x} dx d\tau$$

$$= \left\langle e, \frac{\partial E}{\partial t} \right\rangle + \left\langle \frac{\partial E}{\partial x}, \frac{\partial e}{\partial x} \right\rangle \left\{ 1 + \int_0^t \int_0^1 \left(\frac{\partial \tilde{u}^h}{\partial x} \right)^2 dx d\tau \right.$$

$$\left. + \int_0^t \int_0^1 \left(\frac{\partial E}{\partial x} \right)^2 dx d\tau + \int_0^t \int_0^1 2 \frac{\partial \tilde{u}^h}{\partial x} \frac{\partial E}{\partial x} dx d\tau \right\}$$

$$+ \left\langle \frac{\partial \tilde{u}^h}{\partial x}, \frac{\partial e}{\partial x} \right\rangle \left\{ \int_0^t \int_0^1 \left(\frac{\partial E}{\partial x} \right)^2 dx d\tau + \int_0^t \int_0^1 2 \frac{\partial \tilde{u}^h}{\partial x} \frac{\partial E}{\partial x} dx d\tau \right\}$$

$$\leq \left\langle e, \frac{\partial E}{\partial t} \right\rangle + \left\langle \left| \frac{\partial E}{\partial x} \right|, \left| \frac{\partial e}{\partial x} \right| \right\rangle \left\{ 1 + \int_0^t \int_0^1 \left(\frac{\partial \tilde{u}^h}{\partial x} \right)^2 dx d\tau \right.$$

$$+ \int_0^t \int_0^1 \left(\frac{\partial E}{\partial x} \right)^2 dx d\tau + \frac{1}{\epsilon_1} \int_0^t \int_0^1 \left(\frac{\partial \tilde{u}^h}{\partial x} \right)^2 dx d\tau$$

$$\left. + \epsilon_1 \int_0^t \int_0^1 \left(\frac{\partial E}{\partial x} \right)^2 dx d\tau \right\}$$

$$+ \left\langle \left| \frac{\partial \tilde{u}^h}{\partial x} \right|, \left| \frac{\partial e}{\partial x} \right| \right\rangle \left\{ \int_0^t \int_0^1 \left(\frac{\partial E}{\partial x} \right)^2 dx d\tau \right.$$

$$\left. + 2 \left| \sup \left(\frac{\partial \tilde{u}^h}{\quad} \right) \right| \int^t \int^1 \left| \frac{\partial E}{\quad} \right| dx d\tau \right\}$$

$$\leq \left\langle e, \frac{\partial E}{\partial t} \right\rangle + \left\langle \left|\frac{\partial E}{\partial x}\right|, \left|\frac{\partial e}{\partial x}\right| \right\rangle \left\{ 1 + \int_0^t \int_0^1 \left(\frac{\partial \tilde{u}^h}{\partial x}\right)^2 dx d\tau \right.$$

$$+ \int_0^t \int_0^1 \left(\frac{\partial E}{\partial x}\right)^2 dx d\tau + \frac{1}{\epsilon_1} \int_0^t \int_0^1 \left(\frac{\partial \tilde{u}^h}{\partial x}\right)^2 dx d\tau$$

$$\left. +\epsilon_1 \int_0^t \int_0^1 \left(\frac{\partial E}{\partial x}\right)^2 dx d\tau \right\}$$

$$+ \left| \sup_x \left(\frac{\partial \tilde{u}^h}{\partial x}\right) \right| \int_0^1 \left|\frac{\partial e}{\partial x}\right| dx \left\{ \int_0^t \int_0^1 \left(\frac{\partial E}{\partial x}\right)^2 dx d\tau \right.$$

$$\left. +2 \left| \sup_{x,t} \left(\frac{\partial \tilde{u}^h}{\partial x}\right) \right| \int_0^t \int_0^1 \left|\frac{\partial E}{\partial x}\right| dx d\tau \right\},$$

where $\epsilon_1 > 0$ comes from Schwarz's inequality.

Now incorporate these estimates into (3.137) to have

$$\frac{1}{2}\frac{d}{dt}\|e\|^2 + \left\|\frac{\partial e}{\partial x}\right\|^2 + \frac{1}{4}\frac{d\varphi^2(t)}{dt} \leq \left\langle e, \frac{\partial E}{\partial t} \right\rangle$$

$$+ \left\langle \left|\frac{\partial E}{\partial x}\right|, \left|\frac{\partial e}{\partial x}\right| \right\rangle \left\{ 1 + \int_0^t \int_0^1 \left(\frac{\partial \tilde{u}^h}{\partial x}\right)^2 dx d\tau \right.$$

$$+ \int_0^t \int_0^1 \left(\frac{\partial E}{\partial x}\right)^2 dx d\tau + \frac{1}{\epsilon_1} \int_0^t \int_0^1 \left(\frac{\partial \tilde{u}^h}{\partial x}\right)^2 dx d\tau$$

$$\left. +\epsilon_1 \int_0^t \int_0^1 \left(\frac{\partial E}{\partial x}\right)^2 dx d\tau \right\}$$

$$+\left|\sup_x\left(\frac{\partial\tilde{u}^h}{\partial x}\right)\right|\int_0^1\left|\frac{\partial e}{\partial x}\right|dx\left\{\int_0^t\int_0^1\left(\frac{\partial E}{\partial x}\right)^2 dxd\tau\right.$$

$$\left.+2\left|\sup_{x,t}\left(\frac{\partial\tilde{u}^h}{\partial x}\right)\right|\int_0^t\int_0^1\left|\frac{\partial E}{\partial x}\right|dxd\tau\right\}.$$

Integrate with respect to t, we have

$$\frac{1}{2}||e(\cdot,T)||^2+\int_0^T\left\|\frac{\partial e}{\partial x}\right\|^2 dt+\frac{1}{4}\varphi^2(T)\leq\frac{1}{2}||e(\cdot,0)||^2$$

$$+\int_0^T\int_0^1 e\frac{\partial E}{\partial t}dxdt$$

$$+\int_0^T\int_0^1\left|\frac{\partial E}{\partial x}\frac{\partial e}{\partial x}\right|dx\left\{1+(1+\epsilon_1)\int_0^t\int_0^1\left(\frac{\partial E}{\partial x}\right)^2 dxd\tau\right.$$

$$\left.+\left(1+\frac{1}{\epsilon_1}\right)\int_0^t\int_0^1\left(\frac{\partial\tilde{u}^h}{\partial x}\right)^2 dxd\tau\right\}dt$$

$$+\int_0^T\left|\sup_x\left(\frac{\partial\tilde{u}^h}{\partial x}\right)\right|\int_0^1\left|\frac{\partial e}{\partial x}\right|dx\left\{\int_0^t\int_0^1\left(\frac{\partial E}{\partial x}\right)^2 dxd\tau\right.$$

$$\left.+2\left|\sup_{x,t}\left(\frac{\partial\tilde{u}^h}{\partial x}\right)\right|\int_0^t\int_0^1\left|\frac{\partial E}{\partial x}\right|dxd\tau\right\}dt. \tag{3.140}$$

Note that the first and third terms on the left are nonnegative and can be dropped. We also use Schwarz's inequality on the right-hand side for these two terms:

$$\int_0^T\int_0^1 e\frac{\partial E}{\partial t}dxdt\leq\frac{1}{2}\epsilon_2|||e|||^2+\frac{1}{2\epsilon_2}\left|\left|\left|\frac{\partial E}{\partial t}\right|\right|\right|^2,$$

$$\int_0^T \int_0^1 \left| \frac{\partial E}{\partial x} \frac{\partial e}{\partial x} \right| dx \le \frac{1}{2}\epsilon_3 \left|\left|\left| \frac{\partial E}{\partial x} \right|\right|\right|^2 + \frac{1}{2\epsilon_3} \left|\left|\left| \frac{\partial e}{\partial x} \right|\right|\right|^2 .$$

We can estimate the last term on the right by estimating the term in parenthesis and then take it outside the time integral

$$\begin{aligned} I &\equiv \int_0^t \int_0^1 \left(\frac{\partial E}{\partial x} \right)^2 dx d\tau + 2 \left| \sup_{x,t} \left(\frac{\partial \tilde{u}^h}{\partial x} \right) \right| \int_0^t \int_0^1 \left| \frac{\partial E}{\partial x} \right| dx d\tau \\ &\le \left|\left|\left| \frac{\partial E}{\partial x} \right|\right|\right|^2 + 2 \left| \sup_{x,t} \left(\frac{\partial \tilde{u}^h}{\partial x} \right) \right| \int_0^T \int_0^1 \left| \frac{\partial E}{\partial x} \right| dx d\tau \\ &= \left|\left|\left| \frac{\partial E}{\partial x} \right|\right|\right|^2 + 2 \left| \sup_{x,t} \left(\frac{\partial \tilde{u}^h}{\partial x} \right) \right| \left[\!\left] \frac{\partial E}{\partial x} \right[\!\right] . \end{aligned}$$

Therefore, the last term in (3.140) becomes

$$\begin{aligned} & I \int_0^T \left| \sup_x \left(\frac{\partial \tilde{u}^h}{\partial x} \right) \right| \int_0^1 \left| \frac{\partial e}{\partial x} \right| dx dt \\ & \le I \sqrt{\int_0^T \left| \sup_x \left(\frac{\partial \tilde{u}^h}{\partial x} \right) \right|^2 dt} \sqrt{\int_0^T \left(\int_0^1 \left| \frac{\partial e}{\partial x} \right| dx \right)^2 dt} \\ & \le I \sqrt{\int_0^T \left| \sup_x \left(\frac{\partial \tilde{u}^h}{\partial x} \right) \right|^2 dt} \left|\left|\left| \frac{\partial e}{\partial x} \right|\right|\right| \\ & \le \frac{\epsilon_4}{2} I^2 \int_0^T \left| \sup_x \left(\frac{\partial \tilde{u}^h}{\partial x} \right) \right|^2 dt + \frac{1}{2\epsilon_4} \left|\left|\left| \frac{\partial e}{\partial x} \right|\right|\right|^2 . \end{aligned}$$

Therefore, (3.140) gives

$$\left|\left|\left|\frac{\partial e}{\partial x}\right|\right|\right|^2 \le \frac{1}{2}||e(\cdot,0)||^2 + \frac{\epsilon_2}{2}|||e|||^2 + \frac{1}{2\epsilon_2}\left|\left|\left|\frac{\partial E}{\partial t}\right|\right|\right|^2$$

$$+M_1\left(\frac{\epsilon_3}{2}\left|\left|\left|\frac{\partial E}{\partial x}\right|\right|\right|^2 + \frac{1}{2\epsilon_3}\left|\left|\left|\frac{\partial e}{\partial x}\right|\right|\right|^2\right)$$

$$+(1+\epsilon_1)\left|\left|\left|\frac{\partial E}{\partial x}\right|\right|\right|^2\left(\frac{\epsilon_3}{2}\left|\left|\left|\frac{\partial E}{\partial x}\right|\right|\right|^2 + \frac{1}{2\epsilon_3}\left|\left|\left|\frac{\partial e}{\partial x}\right|\right|\right|^2\right)$$

$$+\frac{\epsilon_4}{2}I^2\int_0^T\left|\sup_x\left(\frac{\partial \tilde{u}^h}{\partial x}\right)\right|^2 dt + \frac{1}{2\epsilon_4}\left|\left|\left|\frac{\partial e}{\partial x}\right|\right|\right|^2,$$

where

$$M_1 \equiv 1 + \left(1+\frac{1}{\epsilon_1}\right)\int_0^T\int_0^1\left(\frac{\partial \tilde{u}^h}{\partial x}\right)^2 dx d\tau.$$

Combining terms we have

$$\left[1 - \frac{M_1}{2\epsilon_3} - \frac{1}{2\epsilon_4} - \frac{1+\epsilon_1}{2\epsilon_3}\left|\left|\left|\frac{\partial E}{\partial x}\right|\right|\right|^2\right]\left|\left|\left|\frac{\partial e}{\partial x}\right|\right|\right|^2 - \frac{\epsilon_2}{2}|||e|||^2$$

$$\le \frac{1}{2}||e(\cdot,0)||^2 + \frac{1}{2\epsilon_2}\left|\left|\left|\frac{\partial E}{\partial t}\right|\right|\right|^2$$

$$+\left[\frac{M_1\epsilon_3}{2} + \frac{(1+\epsilon_1)\epsilon_3}{2}\left|\left|\left|\frac{\partial E}{\partial x}\right|\right|\right|^2\right]\left|\left|\left|\frac{\partial E}{\partial x}\right|\right|\right|^2$$

$$+\frac{\epsilon_4}{2}\int_0^T\left|\sup_x\left(\frac{\partial \tilde{u}^h}{\partial x}\right)\right|^2 dt\left(\left|\left|\left|\frac{\partial E}{\partial x}\right|\right|\right|^2 + 2\left|\sup_{x,t}\left(\frac{\partial \tilde{u}^h}{\partial x}\right)\right|\left[\!\left]\frac{\partial E}{\partial x}\right[\!\right]\right)^2.$$

Now we use Poincaré's inequality

$$\left|\left|\frac{\partial e}{\partial x}\right|\right| \geq C_p ||e||$$

to show that

$$\left|\left|\left|\frac{\partial e}{\partial x}\right|\right|\right| \geq C_p |||e|||$$

for possibly different constant C_p.

We have

$$\left[1 - \frac{M_1}{2\epsilon_3} - \frac{1}{2\epsilon_4} - \frac{1+\epsilon_1}{2\epsilon_3}\left|\left|\left|\frac{\partial E}{\partial x}\right|\right|\right|^2 - \frac{\epsilon_2}{2C_p^2}\right]\left|\left|\left|\frac{\partial e}{\partial x}\right|\right|\right|^2$$

$$\leq \frac{1}{2}||e(\cdot,0)||^2 + \frac{1}{2\epsilon_2}\left|\left|\left|\frac{\partial E}{\partial t}\right|\right|\right|^2$$

$$+\left[\frac{M_1\epsilon_3}{2} + \frac{(1+\epsilon_1)\epsilon_3}{2}\left|\left|\left|\frac{\partial E}{\partial x}\right|\right|\right|^2\right]\left|\left|\left|\frac{\partial E}{\partial x}\right|\right|\right|^2 \qquad (3.141)$$

$$+\frac{\epsilon_4}{2}\int_0^T \left|\sup_x\left(\frac{\partial \tilde{u}^h}{\partial x}\right)\right|^2 dt \left(\left|\left|\left|\frac{\partial E}{\partial x}\right|\right|\right|^2\right.$$

$$\left. +2\left|\sup_{x,t}\left(\frac{\partial \tilde{u}^h}{\partial x}\right)\right| \left[\!\left]\frac{\partial E}{\partial x}\right[\!\right]\right)^2.$$

Note that we can choose ϵ_i, $i = 1, \ldots, 4$ so that the coefficient of the term on the left-hand side of (3.141)

$$M_2 \equiv 1 - \frac{M_1}{2\epsilon_3} - \frac{1}{2\epsilon_4} - \frac{1+\epsilon_1}{2\epsilon_3}\left|\left|\left|\frac{\partial E}{\partial x}\right|\right|\right|^2 - \frac{\epsilon_2}{2C_p^2}$$

is a positive constant. Recall that

$$|||e|||_1^2 = |||e|||^2 + \left|\left|\left|\frac{\partial e}{\partial x}\right|\right|\right|^2 \leq \left(1 + \frac{1}{C_p^2}\right) \left|\left|\left|\frac{\partial e}{\partial x}\right|\right|\right|^2 ,$$

so using (3.141), we get

$$|||e|||_1^2 \leq c_1^2 \left\{ \frac{1}{2}||e(\cdot,0)||^2 + \frac{1}{2\epsilon_2} \left|\left|\left|\frac{\partial E}{\partial t}\right|\right|\right|^2 \right.$$

$$+ \left[\frac{M_1\epsilon_3}{2} + \frac{(1+\epsilon_1)\epsilon_3}{2} \left|\left|\left|\frac{\partial E}{\partial x}\right|\right|\right|^2\right] \left|\left|\left|\frac{\partial E}{\partial x}\right|\right|\right|^2$$

$$\left. +c_2 \left(\left|\left|\left|\frac{\partial E}{\partial x}\right|\right|\right|^2 + 2 \left|\sup_{x,t}\left(\frac{\partial \tilde{u}^h}{\partial x}\right)\right| \left[\!\left[\frac{\partial E}{\partial x}\right]\!\right] \right)^2 \right\} ,$$

where

$$c_1 = \sqrt{\frac{\left(1 + \frac{1}{C_p^2}\right)}{M_2}}$$

and

$$c_2 = \frac{\epsilon_4}{2} \int_0^T \left|\sup_x \left(\frac{\partial \tilde{u}^h}{\partial x}\right)\right|^2 dt\,.$$

Since E is the interpolation error, from (3.131) we have

$$|||E||| \leq C_0 h^j |||U||| ,$$

$$\left|\left|\left|\frac{\partial E}{\partial x}\right|\right|\right| \leq C_1 h^{j-1} |||U||| ,$$

$$\left|\left|\left|\frac{\partial E}{\partial t}\right|\right|\right| \leq C_2 h^j \left|\left|\left|\frac{\partial U}{\partial t}\right|\right|\right| ,$$

$$\left[\!\left]\frac{\partial E}{\partial x}\right[\!\right] \leq C_3 h^{j-1} [\!] U [\!] \,,$$

which yield

$$\begin{aligned}|||e|||_1 \leq h^{j-1} c_1 \Bigg\{ & \frac{1}{2} h^2 ||U_0||^2 + c_3 h^2 \left|\left|\left|\frac{\partial U}{\partial t}\right|\right|\right|^2 \\ & + \left[c_4 + c_5 h^{2(j-1)} |||U|||^2\right] c_7 |||U|||^2 \\ & + c_2 \left(c_7 h^{j-1} |||U|||^2 + c_6 [\!] U [\!] \right)^2 \Bigg\}^{1/2} ,\end{aligned} \tag{3.142}$$

where

$$\begin{aligned} & c_3 = \frac{1}{2\epsilon_2} C_2^2 \,, \quad c_4 = \frac{M_1 \epsilon_3}{2} \,, \quad c_5 = \frac{(1+\epsilon_1)\epsilon_3}{2} C_1^2 \,, \\ & c_6 = 2 \left| \sup_{x,t} \left(\frac{\partial \tilde{u}^h}{\partial x} \right) \right| C_3 \,, \quad c_7 = C_1^2 \,. \end{aligned} \tag{3.143}$$

Using the triangle inequality we get

$$|||U - u^h|||_1 = |||U - \bar{u}^h + \bar{u}^h - u^h|||_1 \leq |||E|||_1 + |||e|||_1 .$$

Therefore,

$$\begin{aligned}|||U - u^h|||_1^2 \leq \; & c_8^2 h^{2(j-1)} |||U|||^2 \\ & + h^{2(j-1)} c_1^2 \Bigg\{ \frac{1}{2} h^2 ||U_0||^2 + c_3 h^2 \left|\left|\left|\frac{\partial U}{\partial t}\right|\right|\right|^2 \\ & + \left[c_4 + c_5 h^{2(j-1)} |||U|||^2\right] c_7 |||U|||^2 \\ & + c_2 \left(c_7 h^{j-1} |||U|||^2 + c_6 [\!] U [\!] \right)^2 \Bigg\}\end{aligned}$$

or

$$|||U - u^h|||_1^2 \leq h^{2(j-1)} c_1^2 \left\{ \frac{1}{2} h^2 ||U_0||^2 \right.$$

$$+c_3 h^2 \left|\left|\left| \frac{\partial U}{\partial t} \right|\right|\right|^2 + \left[c_9 + c_5 h^{2(j-1)} |||U|||^2 \right] c_7 |||U|||^2 \tag{3.144}$$

$$\left. +c_2 \left(c_7 h^{j-1} |||U|||^2 + c_6 [] U [] \right)^2 \right\},$$

where $c_9 = c_4 + \frac{c_8^2}{c_1^2 c_7}$. From (3.144), by denoting $C = \max_{i=1,2,\ldots,9} \{c_i\}$, we complete proof of Theorem 3.6.

The solution of a system (3.132), (3.133) will be discussed in Chapter 4. We have carried out various numerical experiments. These experiments agree with theoretical research (see section 4.3 of Chapter 4).

3.8 Galerkin's method for Model III

In this section we construct and study the Galerkin's approximation for problem (3.68)-(3.70).

3.8.1 Variational formulation

Let us note once again that one of the ingredients of the finite element method is a variational formulation of the problem. To provide this variational formulation, let us denote by H the linear space of functions u satisfying the boundary conditions in (3.69) and

$$||u(\cdot, t)||_1 < \infty,$$

where the norm is as before, see (3.128). The variational formulation of the problem can be stated as follows:

Find a function $U(x,t) \in H$ for which

$$\left\langle V, \frac{\partial U}{\partial t} \right\rangle + \int_0^t a(t-\tau) \left\langle \sigma \left(\frac{\partial U(x,\tau)}{\partial x} \right), \frac{\partial V(x,t)}{\partial x} \right\rangle d\tau$$

$$= \langle V, f \rangle, \; \forall V \in H, \tag{3.145}$$

and

$$\langle V, U(x,0) \rangle = \langle V, U_0(x) \rangle, \; \forall V \in H, \tag{3.146}$$

where the inner product is as defined before.

To approximate the solution of (3.145), (3.146) we require that U and V lie in a finite dimensional subspace H^h of H for each t (see, e.g., [375]). The approximation property in H^h (3.131) can be readily verified for finite element spaces (see, e.g., [439] and section 3.6).

Approximate Problem

The approximation $u^h \in H^h$ to U is defined by the following variational analogue of (3.145), (3.146):
Find a $u^h \in H^h$ such that

$$\left\langle v^h, \frac{\partial u^h(x,t)}{\partial t} \right\rangle$$

$$+ \int_0^t a(t-\tau) \left\langle \sigma \left(\frac{\partial u^h(x,\tau)}{\partial x} \right), \frac{\partial v^h(x,t)}{\partial x} \right\rangle d\tau \tag{3.147}$$

$$= \left\langle f, v^h \right\rangle, \qquad \forall v^h \in H^h,$$

and

$$\left\langle v^h, u^h(x,0) \right\rangle = \left\langle v^h, U_0(x) \right\rangle, \qquad \forall v^h \in H^h. \tag{3.148}$$

Once a basis has been selected for H^h, (3.147), (3.148) are equivalent to a set of N integro-differential equations, where N is dimension of H^h.

3.8.2 Error estimates

.

In this section we shall estimate the error in the finite element approximation using the norm (3.134). We will also use the following norm

$$[]\xi[]_r = \left\{ \int_0^1 \sum_{i=0}^{r} \left[\sup_{0\le t\le T} \left| \frac{\partial^i \xi(x,t)}{\partial x^i} \right| \right]^2 dx \right\}^{1/2}. \tag{3.149}$$

Whenever $r = 0$ we will omit subscript.

This error estimate depends upon the assumption that there exist $\beta > 0$ such that for any function $\omega(x,t)$ we have

$$\begin{aligned} &\int_0^T \int_0^t a(t-\tau) \int_0^1 \sigma'(\omega(x,\tau))\xi(x,\tau)\xi(x,t)dxd\tau dt \\ &\ge \frac{\beta A^2}{a(0)} \int_0^1 \left\{ \int_0^T |\xi(x,\tau)|\, d\tau \right\}^2 dx, \end{aligned} \tag{3.150}$$

where $A = \inf_{0\le t\le T} a(t)$. This assumption is analogous to the inequality for positive definite functions proved by Staffans [438]. It becomes the consequence of Lemma 4.1 of [438] if $\sigma'(\omega(x,\tau)) \equiv \beta$.

Theorem 3.7 *The error in the finite element approximation u^h generated by (3.147), (3.148) satisfies the relation*

$$\begin{aligned} |||U - u^h|||_1 \le h^{r-1} \Bigg\{ & C_2^2 h^2 ||U_0||_r^2 + \frac{C_0^2 h^2}{\epsilon} \left[\!\left] \frac{\partial U}{\partial t} \right[\!\right]_r^2 \\ & + \frac{\gamma^2 M^2 C_1^2}{\delta} |||U|||_r^2 \Bigg\}^{1/2} + C_1 h^{r-1} |||U|||_r. \end{aligned} \tag{3.151}$$

Proof. Subtracting (3.147) from (3.145) with v^h instead of v, we obtain

$$\begin{aligned}&\left\langle v^h, \frac{\partial u^h}{\partial t}\right\rangle\\&+\int_0^t a(t-\tau)\left\langle \sigma\left(\frac{\partial u^h(x,\tau)}{\partial x}\right), \frac{\partial v^h(x,t)}{\partial x}\right\rangle d\tau\\&=\left\langle v^h, \frac{\partial U}{\partial t}\right\rangle\\&+\int_0^t a(t-\tau)\left\langle \sigma\left(\frac{\partial U(x,\tau)}{\partial x}\right), \frac{\partial v^h(x,t)}{\partial x}\right\rangle d\tau,\\&\qquad \forall v^h \in H^h.\end{aligned} \tag{3.152}$$

Let $\tilde{u}^h$ be any function in H^h, then

$$\begin{aligned}&\left\langle v^h, \frac{\partial (u^h-\tilde{u}^h)}{\partial t}\right\rangle\\&+\int_0^t a(t-\tau)\left\langle \sigma\left(\frac{\partial u^h(x,\tau)}{\partial x}\right)-\sigma\left(\frac{\partial \tilde{u}^h(x,\tau)}{\partial x}\right), \frac{\partial v^h(x,t)}{\partial x}\right\rangle d\tau\\&=\left\langle v^h, \frac{\partial (U-\tilde{u}^h)}{\partial t}\right\rangle\\&+\int_0^t a(t-\tau)\left\langle \sigma\left(\frac{\partial U(x,\tau)}{\partial x}\right)-\sigma\left(\frac{\partial \tilde{u}^h(x,\tau)}{\partial x}\right), \frac{\partial v^h(x,t)}{\partial x}\right\rangle d\tau,\\&\qquad \forall v^h \in H^h,\end{aligned} \tag{3.153}$$

Let $e(x,t)$ and $E(x,t)$ be defined as in (3.136). Since $e \in H^h$ we can let $v^h = e$. Using the mean value theorem and (2.63)

$$\begin{aligned}
&\frac{1}{2}\frac{d}{dt}||e(\cdot,t)||^2 \\
&+\int_0^t a(t-\tau)\left\langle \sigma'(\xi(x,\tau))\frac{\partial e(x,\tau)}{\partial x}, \frac{\partial e(x,t)}{\partial x}\right\rangle d\tau \\
&\leq \left\langle e(x,t), \frac{\partial E(x,t)}{\partial t}\right\rangle \\
&+\gamma\left|\int_0^t a(t-\tau)\int_0^1 \frac{\partial e(x,t)}{\partial x}\frac{\partial E(x,\tau)}{\partial x}dxd\tau\right|.
\end{aligned} \tag{3.154}$$

Using Schwarz's inequality on the last integral and integrating with respect to t, we have

$$\begin{aligned}
&||e(\cdot,T)||^2 - ||e(\cdot,0||^2 \\
&+2\int_0^T dt\int_0^t d\tau a(t-\tau)\int_0^1 dx\sigma'(\xi(x,\tau))\frac{\partial e(x,\tau)}{\partial x}\frac{\partial e(x,t)}{\partial x} \\
&\leq \int_0^T dt\int_0^1 e(x,t)\frac{\partial E(x,t)}{\partial t}dx \\
&+2\gamma\int_0^1 dx\int_0^T dt\left|\frac{\partial e(x,t)}{\partial x}\right|\left\{\int_0^t a^2(t-\tau)d\tau\right\}^{1/2} \\
&\times\left\{\int_0^t\left(\frac{\partial E(x,\tau)}{\partial x}\right)^2 d\tau\right\}^{1/2}.
\end{aligned} \tag{3.155}$$

Using Schwarz's inequality on the integrals on the right and

(3.150) it follows that

$$||e(\cdot,T)||^2 + \frac{\beta A^2}{a(0)} \int_0^1 dx \left\{ \int_0^T \left| \frac{\partial e(x,\tau)}{\partial x} \right| d\tau \right\}^2$$

$$\leq ||e(\cdot,0||^2 + 2 \int_0^1 \left\{ \int_0^T |e(x,t)|dt \right\} \sup_t \left| \frac{\partial E(x,t)}{\partial t} \right|$$

$$+2\gamma \int_0^1 dx \left\{ \int_0^T \left| \frac{\partial e(x,t)}{\partial x} \right| dt \right\} \sup_t \left\{ \int_0^t a^2(t-s)ds \right\}^{1/2}$$

$$\times \left\{ \int_0^T \left(\frac{\partial E(x,\tau)}{\partial x} \right)^2 d\tau \right\}^{1/2}. \tag{3.156}$$

Using Schwarz's inequality again and noting that $||e(\cdot,T)||^2 \geq 0$, one has

$$\frac{\beta A^2}{a(0)} \left|\left|\left| \frac{\partial e}{\partial x} \right|\right|\right|^2 \leq ||e(\cdot,0||^2$$

$$+2 \left\{ \int_0^1 dx \left\{ \int_0^T |e(x,t)|dt \right\}^2 \right\}^{1/2}$$

$$\times \left\{ \int_0^1 dx \sup_t \left| \frac{\partial E(x,t)}{\partial t} \right| \right\}^{1/2} \tag{3.157}$$

$$+2\gamma M \left\{ \int_0^1 dx \left\{ \int_0^T \left| \frac{\partial e(x,t)}{\partial x} \right| dt \right\}^2 \right\}^{1/2}$$

$$\times \left\{ \int_0^1 dx \int_0^T \left(\frac{\partial E(x,s)}{\partial x} \right)^2 ds \right\}^{1/2},$$

where

$$M = \sup_t \left\{ \int_0^t a^2(t-\tau)d\tau \right\}^{1/2}.$$

Since $2ab \leq \epsilon a^2 + b^2/\epsilon$ for $\epsilon > 0$, one has after combining like terms

$$\left(\frac{\beta A^2}{a(0)} - \delta \right) \left|\left|\left| \frac{\partial e}{\partial x} \right|\right|\right|^2 - \epsilon |||e|||^2$$
$$\leq ||e(\cdot, 0)||^2 + \frac{1}{\epsilon} \left[\!\left] \frac{\partial E}{\partial t} \right[\!\right]^2 + \frac{\gamma^2 M^2}{\delta} \left|\left|\left| \frac{\partial E}{\partial x} \right|\right|\right|_r^2. \tag{3.158}$$

Using Poincaré's inequality

$$\left|\left| \frac{\partial e}{\partial x} \right|\right| \geq C_p ||e||,$$

one can show that

$$\left|\left|\left| \frac{\partial e}{\partial x} \right|\right|\right| \geq C_p |||e|||$$

for possibly different constant. Let $\delta > 0$ and $\epsilon > 0$ be chosen so that

$$\left(\frac{\beta A^2}{a(0)} - \delta - \frac{\epsilon}{C_p^2} \right) = 1 + \frac{1}{C_p^2}$$

and note that

$$|||e|||_1^2 = \left|\left|\left| \frac{\partial e}{\partial x} \right|\right|\right|^2 + |||e|||^2 \geq \left(1 + \frac{1}{C_p^2} \right) \left|\left| \frac{\partial e}{\partial x} \right|\right|^2.$$

Therefore,

$$|||e|||_1^2 \leq ||e(\cdot, 0)||^2 + \frac{1}{\epsilon} \left[\!\left] \frac{\partial E}{\partial t} \right[\!\right]^2 + \frac{\gamma^2 M^2}{\delta} \left|\left|\left| \frac{\partial E}{\partial x} \right|\right|\right|_r^2.$$

Using the approximation property and noting that the error in the initial condition satisfies

$$||e(\cdot, 0)|| \leq C_2 h^r ||U_0||_r,$$

one obtains

$$|||e|||_1^2 \leq C_2^2 h^{2r} ||U_0||^2 + \frac{C_0^2 h^{2r}}{\epsilon} \left[\!\left]\frac{\partial U}{\partial t}\right[\!\right]_r^2$$

$$+\frac{\gamma^2 M^2}{\delta} C_1^2 h^{2(r-1)} |||U|||_r^2.$$

Taking square roots of both sides and using the triangle inequality we obtain the error estimate (3.151).

Thus, Theorem 3.7 has been proved.

3.9 Comments and bibliographical notes

Due to the complexity of the real problems, it is very hard to obtain exact solutions or it is too complex to have practical use. As a result, scientists have to look for numerical methods for these equations.

There are many scientific papers and books on the numerical solution of these problems. Several approximate methods are constructed and investigated for numerical solution of ordinary and partial differential equations, see, for example, [23], [54], [72], [73], [139], [148], [153], [161], [167], [170], [171], [172], [173], [174], [175], [183], [193], [207], [210], [211], [212], [213], [214], [298], [301], [314], [315], [316], [317], [327], [336], [337], [341], [350], [369], [371], [375], [400], [402], [403], [404], [407], [411],

[415], [416], [421], [422], [434], [435], [439], [440], [442], [446], [447], [472], [474], [480], [481], [493], [496], [498], [502], [503], and [504]. For integral equations, see [41], [42], [48], [78], [79], [80], [117], [175], [189], [290], [292], [326], [377], [379], [412] and [482] and for integro-differential equations, see [12], [16], [28], [29], [31], [32], [33], [35], [43], [45], [46], [47], [57], [58], [63], [69], [74], [78], [80], [81], [82], [83], [85], [86], [88], [89], [90], [97], [98], [103], [112], [113], [117], [118], [125], [126], [139], [154], [161], [163], [166], [175], [178], [198], [200], [214], [220], [223], [227], [239], [241], [243], [244], [255], [256], [257], [258], [261], [262], [263], [268], [269], [272], [274], [276], [277], [280], [283], [285], [298], [308], [313], [319], [320], [321], [327], [328], [333], [334], [335], [342], [343], [348], [351], [353], [354], [366], [368], [376], [378], [387], [388], [389], [390], [393], [408], [410], [413], [420], [425], [433], [443], [449], [450], [456], [457], [464], [478], [479], [495], and [497].

Many scientific works are devoted to numerical solution of Maxwell's systems and Maxwell's type systems as well, see [54], [60], [66], [139], [148], [153], [172], [173], [174], [194], [197], [205], [212], [214], [220], [223], [227], [239], [241], [243], [244], [255], [256], [257], [258], [261], [274], [276], [277], [280], [283], [314], [315], [316], [317], [333], [334], [336], [341], [369], [434], [442], [472], [480], [493], [498], [504], and references therein.

Maxwell's nonstationary models with just Joule's term or with thermal conductivity as well, which are the parabolic structures, by using the different approximate methods, are studied in many works (see, for example, [4], [5], [19], [26], [27], [38], [39], [105], [139], [142], [143], [145], [148], [153], [172], [173], [174], [189], [205], [209], [212], [221], [224], [225], [226], [230], [247], [253], [273], [276], [314], [315], [316], [336], [341], [369], [377], [379], [434], [480], [494], [498], [499], and references therein).

In numerous scientific literature Maxwell's nonlinear elliptic-parabolic and stationary models are considered as well. In this direction, for example, in [15] fully discrete numerical methods for a nonlinear elliptic-parabolic system, the time-dependent Joule heating problem, are introduced and analyzed. The equations are discretized in space by a standard finite element method and in time by combinations of rational implicit and explicit multistep schemes. The schemes are linearly implicit in the sense that they require, at each time level, the solution of linear systems of equations. Optimal order error estimates are proved under the assumption of sufficient regularity of solutions. A finite element method is considered for this type model in [153]. The work [106] and several other works are devoted to investigation of such models as well.

Many scientific works are devoted to approximate solution of integral and singular integral equations. The results obtained in investigations of these models play a great role in construction of numerical solutions of partial integro-differential models [41], [42], [48], [78], [80], [157], [215], [292], [326], [398], [412], and [482].

Various approximate methods are investigated for numerical solution of evolution integro-differential models. See, for example, [93], [161], [163], [198], [243], [256], [258], [261], [262], [263], [268], [269], [285], [308], [313], [319], [320], [321], [322], [335], [342], [343], [351], [353], [354], [391], [392], [393], [432], [433], [443], [444], [448], [449], and [450]. For example, in [132], [324], [389], [393], [456], [457], and [479] numerical methods for hyperbolic and parabolic integro-differential equations are studied.

Numerical solution of Kirchhoff and its type models is given in many scientific works as well. See, for example, [9], [40], [61], [102], [331], [390], [408], and references there.

It is well known that many works are devoted to investigation of Galerkin's method for nonlinear models, see [17], [47], [104], [129], [175], [258], [261], [298], [308], [316], [327], [328], [336], [421], [447], [461] and references there. Evolution integro-differential models studied by Galerkin's method in many works as well [47], [86], [103], [175], [258], [261], [308], [319], [325], [327], [328], [342], [354], and [497].

For the (1.12) and (1.13) type integro-differential models Galerkin's method algorithms are constructed and studied in [145] and [219].

A finite element method is very important and many engineering and mathematical models are solved by this method, see, e.g., [54], [88], [97], [98], [118], [153], [161], [170], [171], [205], [211], [258], [261], [262], [263], [313], [314], [315], [317], [320], [321], [327], [328], [341], [354], [369], [375], [388], [402], [403], [421], [432], [439], [447], [449], [479], [498], [502], [503], and [504]. Many publications are devoted to the theory and application of a finite element method, see, e.g., [43], [71], [327], [369], [402], [403], [416], [439], [447], [498], [502], [503], and [504]. Many diffusion models including the Maxwell's-type systems are solved by this method (see, [498], [504], and references there).

For the (2.22) and (2.23) type integro-differential models Galerkin's finite element algorithms are constructed and studied in [96], [258], [261], [281], [423], [424], and [501].

Semi-discrete schemes are very often used for investigation of differential and integro differential models too. For integro-differential equations such researches are given in [45], [46], [69], [268], [269], [327], [333], [334], [351], [433], and in a number of other works as well.

For the (2.22) and (2.23) type models semi-discrete schemes are constructed and studied in [30], [139], [145], [248], [249], [250], [251], [252], [253], [276], [282], [284], [333], and [334].

One of the important and very often used methods of numerical solution of differential and integro-differential models is the finite difference method (see [112], [113], [118], [126], [167], [183], [243], [255], [256], [327], [335], [343], [350], [407], [411], [415], [435], [440], [442], [443], [446], and [457]).

Many articles were devoted to the theory and application of a finite difference method (see [167], [183], [298], [327], [407], [411], [415], [435], [440], and [446]). Many diffusion models including the Maxwell's type systems are studied by this method (see [23], [139], [142], [145], [148], [172], [173], [174], [212], [214], [225], [226], [276], and [442]).

For the (2.22) and (2.23) type models finite difference schemes are constructed and studied, for example, in [139], [142], [145], [225], [226], [248], [276], [280], and [318].

Many other methods are also used for numerical solution of differential and integro-differential equations (see [1], [2], [3], [8], [16], [154], [163], [178], [198], [200], [201], [202], [268], [269], [272], [285], [308], [313], [353], [354], [366], [393], [410], [420], [449], and [478]). For example, the paper [16] suggests four different methods to solve nonlinear integro-differential equations, namely, He's variational iteration method, Adomian decomposition method, He's homotopy perturbation method, and differential transform method. The paper [200] deals with Adomian decomposition method. The paper [154] aims to introduce a comparison of Adomian decomposition method and wavelet-Galerkin method for the solution of integro-differential equations. It is noted in [154] that from the computational viewpoint, the comparison in the considered examples shows that the Adomian decomposition method is efficient and easy to use. Wavelet discretizations of parabolic integro-differential equations is studied in [393] as well.

For numerical solution of corresponding problems for Model III many different methods of approximate solution are used (see, for example, [20], [80], [125], [401]).

Now we give more concrete bibliographic references and short comments on separate sections of Chapter 3.

3.1 Semi-discrete scheme for Model I

Semi-discrete schemes with respect to space variable x for one-dimensional integro-differential (2.22) type equations are constructed and studied in the works [139], [145], [220], [248] and [252]. Some of these type discrete analogues are given in [333], and [334] as well.

Let us note that such investigations for one-dimensional integro-differential (2.22) type systems with two-component magnetic field are also given in the works [247], [250], [253], [274], [276], [282], and [284].

3.2 Finite difference scheme for Model I

Finite difference schemes for one-dimensional integro-differential (2.22) type equations are given in the works [139], [145], [220], [227], [239], [243], [244], [248], [333], and [334]. Such questions for an initial-boundary value problem with mixed boundary conditions for a (2.22) type integro-differential equation are studied in [280].

In [35] such investigations are done for (2.22) type equations with source term.

Let us note that numerical solution of even one-dimensional integro-differential (2.22) type systems with two-component vector of the magnetic field has its complexities. Now investigations in this direction are continuing, see, for example, [241], [247], [253], [255], [274], [276], and [283].

3.3 Semi-discrete scheme for Model II

Semi-discrete scheme for one-dimensional integro-differential (2.23) type averaged equations are given in the works [145], [249], [251], [256], and [282].

Let us note that such investigations for one-dimensional integro-differential (2.23) type averaged systems with two-component magnetic field are also given in the works [223] and [276].

3.4 Finite difference scheme for Model II

Finite difference schemes for one-dimensional integro-differential (2.23) type averaged equations are given in the works [31], [32], [145], [239], [243], and [256]. In [28] such investigation is done for (2.23) type above-mentioned equations with source term.

Let us note that such investigations for one-dimensional integro-differential (2.23) type averaged systems with two-component magnetic field are also given in the works [33], [223], [241], [247], [253], [255], [257], and [276].

In [29] and [30] such investigation is done for (2.23) type systems with source terms.

3.5 Descrete analogues of Model III

In [378] difference schemes to obtain a numerical solution to the integro-differential equation describing heat flow in materials with memory are studied.

The issue regarding convergence of finite difference as well as semi-discrete schemes, as we already mentioned in the best of authors' knowledge, is still open.

3.6 Galerkin's method for Model I

For the (2.22) type integro-differential models with $a(S) = 1 + S$ Galerkin's finite element algorithm are constructed and studied in [188].

Note that such investigations for multidimensional integro-differential (2.22) are also given in [137], [139], [140], [141], [219].

3.7 Galerkin's method for Model II

For the (2.23) type averaged integro-differential models with $a(S) = 1 + S$ Galerkin's finite element algorithm is constructed and studied in: [96], [258], [261], [281], [423], [424], and [501].

Note that such investigations for one-dimensional integro-differential (2.23) type averaged systems with two-component magnetic field are also given in [259] and [261].

3.8 Galerkin's method for Model III

In [376] Galerkin's method to obtain a numerical solution to the integro-differential equation describing heat flow in materials with memory is studied.

Chapter 4

Numerical Realization of the Discrete Analogous for Models I-III

Abstract

This chapter consists of five sections. The questions of the numerical realizations of algorithms investigated in the third chapter for Models I-III are discussed in this chapter. The approximate models solution of the previously considered approximate models are given. Several examples for numerical solution of Models I-III are presented. Results of various approximate models with graphical illustrations and their analysis are also given in this chapter. In all cases the coincidence of theoretical results given in previous chapters with graphical illustrations and tables is pointed out. At the end of the chapter, comments and bibliographical notes is given for numerical solution of models considered in the book.

Key words: Algorithms of numerical solution, numerical realization, finite-difference and Galerkin's methods, graphics,

Numerical Solutions of Three Classes of Nonlinear Parabolic Integro-Differential Equation.
http://dx.doi.org/10.1016/B978-0-12-804628-9.50004-1

tables.

4.1 Finite difference solution of Model I

In the present section questions of the numerical realization of the finite difference scheme (3.19)-(3.21) investigated in section 3.2 of Chapter 3 for initial-boundary value problem (3.1)-(3.3) are considered.

We now comment on the numerical implementation of the discrete problem (3.19)-(3.21). Note that (3.19) can be rewritten as:

$$\frac{u_i^{j+1} - u_i^j}{\tau} -$$

$$\frac{1}{h}\left\{\left[1 + \tau\sum_{k=1}^{j+1}\left(\frac{u_{i+1}^k - u_i^k}{h}\right)^2\right]\frac{u_{i+1}^{j+1} - u_i^{j+1}}{h} -\right.$$

$$\left.\left[1 + \tau\sum_{k=1}^{j+1}\left(\frac{u_i^k - u_{i-1}^k}{h}\right)^2\right]\frac{u_i^{j+1} - u_{i-1}^{j+1}}{h}\right\} = f_i^j,$$

$$i = 1, \ldots, M-1.$$

Let

$$A_i^\ell = 1 + \tau\sum_{k=1}^{\ell}\left[\left(\frac{u_{i+1}^k - u_i^k}{h}\right)^2\right],$$

$$i = 0, 1, \ldots, M-1,$$

then (3.19) becomes

$$\begin{aligned}
\frac{u_i^{j+1}-u_i^j}{\tau} - \frac{1}{h}\Bigg\{ A_i^{j+1}\frac{u_{i+1}^{j+1}-u_i^{j+1}}{h} - \\
A_{i-1}^{j+1}\frac{u_i^{j+1}-u_{i-1}^{j+1}}{h}\Bigg\} = f_{1,i}^j, \\
i = 1, 2, \ldots, M-1.
\end{aligned} \tag{4.1}$$

System (4.1) can be written in a matrix form

$$\mathbf{H}\left(\mathbf{u}^{j+1}\right) \equiv \mathbf{G}\left(\mathbf{u}^{j+1}\right) - \frac{1}{\tau}\mathbf{u}^j - \mathbf{f}^j = 0.$$

The vector $\mathbf{u}$ containing all the unknowns $u_1, \ldots, u_{M-1}$ at the level indicated. The vector $\mathbf{G}$ is given by

$$\mathbf{G}\left(\mathbf{u}^{j+1}\right) = \mathbf{T}^{j+1}\mathbf{u}^{j+1},$$

where the $(M-1)\times(M-1)$ matrix $\mathbf{T}$ is symmetric and tridiagonal with elements:

$$\mathbf{T}_{rs}^{\ell} = \begin{cases} -\dfrac{1}{h^2}A_{r-1}^{\ell}, & s = r-1, \\ \dfrac{1}{h^2}\left(A_r^{\ell} + A_{r-1}^{\ell}\right), & s = r, \\ -\dfrac{1}{h^2}A_r^{\ell}, & s = r+1, \\ 0, & \text{otherwise}. \end{cases} \tag{4.2}$$

Newton's method for the system is given by

$$\nabla\mathbf{H}\left(\mathbf{u}^{j+1}\right)\Big|^{(n)}\left(\mathbf{u}^{j+1}\Big|^{(n+1)} - \mathbf{u}^{j+1}\Big|^{(n)}\right) = -\mathbf{H}\left(\mathbf{u}^{j+1}\right)\Big|^{(n)}.$$

The elements of the matrix $\nabla \mathbf{H}\left(\mathbf{u}^{j+1}\right)$ require the derivative of A. The elements are

$$\frac{\partial A_{r-1}^{j+1}}{\partial u_s^{j+1}} = -\frac{\tau}{h^2}\frac{\partial}{\partial u_s^{j+1}}\left\{\left(\frac{u_r^{j+1}-u_{r-1}^{j+1}}{h}\right)^2\right\} = \begin{cases} \dfrac{2\tau}{h^3}u_{\bar{x},r}^{j+1}, & s=r-1, \\ -\dfrac{2\tau}{h^3}u_{\bar{x},r}^{j+1}, & s=r, \\ 0, & \text{otherwise,} \end{cases} \tag{4.3}$$

and

$$\frac{\partial A_{r}^{j+1}}{\partial u_s^{j+1}} = -\frac{\tau}{h^2}\frac{\partial}{\partial u_s^{j+1}}\left[\left(\frac{u_{r+1}^{j+1}-u_{r}^{j+1}}{h}\right)^2\right] = \begin{cases} \dfrac{2\tau}{h^3}u_{x,r}^{j+1}, & s=r, \\ -\dfrac{2\tau}{h^3}u_{x,r}^{j+1}, & s=r+1, \\ 0, & \text{otherwise.} \end{cases} \tag{4.4}$$

Combining (4.2)-(4.4) we have

$$\nabla \mathbf{H}\left(\mathbf{u}^{j+1}\right)\Big|_{rs} =$$

$$= \begin{cases} -\dfrac{1}{h^2}A_{r-1}^{j+1} - \dfrac{2\tau}{h^2}\left(u_{\bar{x},r}^{j+1}\right)^2, & s=r-1, \\ \dfrac{1}{\tau} + \dfrac{1}{h^2}\left(A_r^{j+1}+A_{r-1}^{j+1}\right) + \\ \dfrac{2\tau}{h^2}\left(u_{x,r}^{j+1}\right)^2 + \dfrac{2\tau}{h^2}\left(u_{\bar{x},r}^{j+1}\right)^2, & s=r, \\ -\dfrac{1}{h^2}A_r^{j+1} - \dfrac{2\tau}{h^2}\left(u_{x,r}^{j+1}\right)^2, & s=r+1, \\ 0, & \text{otherwise.} \end{cases}$$

Let us state well-known statement (see, for example, [406]).

Theorem 4.1 *Given the nonlinear system of equations*

$$H_i\left(y_1, \ldots, y_{M-1}\right) = 0, \quad i = 1, 2, \ldots, M-1.$$

If H_i are three times continuously differentiable in a region containing the solution $\xi_1, \ldots, \xi_{M-1}$ and the Jacobian does not vanish in that region, then Newton's method converges at least quadratically.

The Jacobian in this theorem is the matrix ∇H computed above. The term $\dfrac{1}{\tau}$ on diagonal ensures that the Jacobian doesn't vanish. The differentiability is guaranteed, since ∇H is quadratic. Newton's method is costly, because the matrix changes at every step of the iteration. One can use modified Newton method (keep the same matrix for several iterations) but the rate of convergence will be slower.

In our first numerical experiment we have chosen the right-hand side of equation (3.1) of problem (3.1)-(3.3) so that the exact solution is given by

$$U(x, t) = x(1-x)\cos t,$$

which satisfies homogeneous boundary conditions (3.2). In this case the right-hand side of (3.1) is

$$f(x,t) = x(x-1)\sin t + \left[\frac{3(1-2x)^2}{2} + 2\right](2t + \sin 2t)\cos t.$$

The parameters used are $M = 100$ which dictates $h = 0.01$. Since the method is implicit we can use $\tau = h$ and we took 100 time steps. In Figures 4.1 and 4.2 we plotted the numerical solution and the exact solutions at $t = 0.5$ (Fig. 4.1) and $t = 1.0$ (Fig. 4.2).

As it is visible from these pictures, the numerical and exact solutions are almost identical.

In the second experiment we have taken zero right-hand side and initial data given by

$$U(x,0) = U_0(x) = x(1-x)\cos(4\pi x).$$

In this case, we know from the asymptotic behavior of the solution of problem (3.1)-(3.3) (see Theorem 2.2) that the solution will decay in time. The parameters M, h, τ are as before. In Figure 4.3 we plotted the initial data and in Figure 4.4 we have the numerical solution at four different times. It is clear that the numerical solution is approaching zero for all x.

The numerical experiments for first kind initial-boundary value problem with nonhomogeneous boundary condition on the right side of lateral boundary (2.66) for problem (3.1), (3.3) were carried out as well. For our next experiment we have taken zero right-hand side and initial data given by

$$U(x,0) = U_0(x) = x(1-x)\cos(4\pi x) + 0.001x.$$

In this case, we know (see (2.68) with $\psi = 0.001$) that the solution will approach the steady state solution, which is $U_s(x) = 0.001x$. The parameters M, h, τ are as before. In Figure 4.5 we plotted the initial data and in Figure 4.6 we have the numerical solution at four different times. It is clear that the numerical solution is approaching steady state solution for all x.

We have experimented with several other initial data for both initial-boundary value problems with homogeneous boundary conditions and with nonhomogeneous boundary condition on the right lateral boundary. In all cases we noticed that numerical solutions are approaching steady state solution as it is dictated by the theoretical researches.

So, the numerical experiments gave the expected results. In all cases numerical solution of our experiments fully agree with the theoretical results given in Chapters 2 and 3.

4.2 Finite difference solution of Model II

In the present section questions of the numerical realization of the finite difference scheme (3.53)-(3.55) investigated in section 3.4 of Chapter 3 for initial-boundary value problem (3.34)-(3.36) are considered.

We now comment on the numerical implementation of the discrete problem (3.53)-(3.55). Note that (3.53) can be rewritten as:

$$\frac{1}{\tau}u_i^{j+1} - A\left(\mathbf{u}^{j+1}\right)\frac{u_{i+1}^{j+1} - 2u_i^{j+1} + u_{i-1}^{j+1}}{h^2} - f_i^j - \frac{1}{\tau}u_i^j = 0,$$

$$i = 1, \ldots, M-1, \tag{4.5}$$

where

$$A\left(\mathbf{u}^{j+1}\right) = 1 + \tau h \sum_{\ell=1}^{M} \sum_{k=1}^{j+1} \left(\frac{u_\ell^k - u_{\ell-1}^k}{h}\right)^2. \tag{4.6}$$

This system (4.5), (4.6) can be written in a matrix form

$$\mathbf{H}\left(\mathbf{u}^{j+1}\right) \equiv \mathbf{G}\left(\mathbf{u}^{j+1}\right) - \frac{1}{\tau}\mathbf{u}^j - \mathbf{f}^j = 0. \tag{4.7}$$

The vector $\mathbf{u}$ contains all the unknowns $u_1, \ldots, u_{M-1}$ at the level indicated. The vector $\mathbf{G}$ is given by

$$\mathbf{G}\left(\mathbf{u}^{j+1}\right) = \mathbf{T}\left(\mathbf{u}^{j+1}\right)\mathbf{u}^{j+1},$$

where the matrix $\mathbf{T}$ is symmetric and tridiagonal with elements

$$\mathbf{T}_{ir} = \begin{cases} \dfrac{1}{\tau} + 2\dfrac{A}{h^2}, & r = i, \\[2ex] -\dfrac{A}{h^2}, & r = i \pm 1. \end{cases}$$

Newton's method for this system is given by

$$\nabla \mathbf{H}\left(\mathbf{u}^{j+1}\right)\Big|^{(n)}\left(\mathbf{u}^{j+1}\Big|^{(n+1)}-\mathbf{u}^{j+1}\Big|^{(n)}\right)=-\mathbf{H}\left(\mathbf{u}^{j+1}\right)\Big|^{(n)}.$$

The elements of the matrix $\nabla \mathbf{H}\left(\mathbf{u}^{j+1}\right)$ require the derivative of A. The elements are

$$\nabla \mathbf{H}\left(\mathbf{u}^{j+1}\right)\Big|_{ir}=$$

$$\begin{cases} \dfrac{1}{\tau}+\dfrac{2}{h^2}A\left(\mathbf{u}^{j+1}\right)-\dfrac{\partial A\left(\mathbf{u}^{j+1}\right)}{\partial u_i^{j+1}}u_{\bar{x}x,i}^{j+1}, & r=i, \\ -u_{\bar{x}x,i}^{j+1}\dfrac{\partial A\left(\mathbf{u}^{j+1}\right)}{\partial u_r^{j+1}}-\dfrac{1}{h^2}A\left(\mathbf{u}^{j+1}\right), & r=i\pm 1, \\ -u_{\bar{x}x,i}^{j+1}\dfrac{\partial A\left(\mathbf{u}^{j+1}\right)}{\partial u_r^{j+1}}, & \text{otherwise,} \end{cases} \tag{4.8}$$

where

$$u_{\bar{x}x,i}^{j+1}=\frac{u_{i+1}^{j+1}-2u_i^{j+1}+u_{i-1}^{j+1}}{h^2}.$$

To evaluate the partial derivatives in (4.8), we use

$$\frac{\partial A}{\partial u_r^{j+1}}=\frac{\partial}{\partial u_r^{j+1}}\left[1+\tau h\sum_{\ell=1}^{M}\sum_{k=1}^{j+1}\left(\frac{u_\ell^k-u_{\ell-1}^k}{h}\right)^2\right]=$$
$$=\frac{\partial}{\partial u_r^{j+1}}\left[C+\tau h\left(\frac{u_r^{j+1}-u_{r-1}^{j+1}}{h}\right)^2+\tau h\left(\frac{u_{r+1}^{j+1}-u_r^{j+1}}{h}\right)^2\right]=$$
$$=2\tau h\frac{u_r^{j+1}-u_{r-1}^{j+1}}{h}\cdot\frac{1}{h}+2\tau h\frac{u_{r+1}^{j+1}-u_r^{j+1}}{h}\cdot\left(-\frac{1}{h}\right)=$$
$$=-2\tau h\frac{u_{r+1}^{j+1}-2u_r^{j+1}+u_{r-1}^{j+1}}{h^2}.$$

Note that we incorporated into the constant C all the terms that are independent of u_r^{j+1}.

So, we have the nonlinear system of equations

$$H_i\left(u_1^{j+1}, \ldots, u_{M-1}^{j+1}\right) = 0, \quad i = 1, 2, \ldots, M-1.$$

As it was already noted in section 4.1, if H_i are three times continuously differentiable in a region containing the solution $\xi_1, \ldots, \xi_{M-1}$ and the Jacobian does not vanish in that region, then Newton's method converges at least quadratically [406].

The Jacobian is the matrix ∇H computed above (4.8). The term $\dfrac{1}{\tau}$ on diagonal ensures that the Jacobian doesn't vanish. The differentiability is guaranteed, since ∇H is quadratic. Therefore, Newton's method, for problem (3.53)-(3.55) converges at least quadratically.

The numerical experiments carried out for the examples in section 4.1 for Model I give similar results as in Figures 4.1-4.6. We have experimented with different forcing terms as well. Those results are not going to be displayed here.

In our first numerical experiment with this model, we have chosen the right-hand side of (3.34) so that the exact solution of problem (3.34)-(3.36) is given by

$$U(x,t) = x(1-x)e^{-x-t}.$$

In this case the right-hand side of (3.34) is

$$\begin{aligned} f(x,t) &= -\left[\tfrac{9}{8} - \tfrac{3}{8}e^{-2} - \left(\tfrac{1}{8} - \tfrac{3}{8}e^{-2}\right)e^{-2t}\right]\left(-4 + 5x - x^2\right)e^{-x-t} \\ &\quad -x(1-x)e^{-x-t}. \end{aligned}$$

In Figures 4.7 and 4.8 we plotted the numerical solution (marked with +) and the exact solution at $t = 0.5$ (Fig. 4.7) and $t = 1.0$ (Fig. 4.8) and it is clear that the two solutions are identical.

In our next experiment we have taken zero right-hand side and initial solution given by

$$U(x, 0) = x(1 - x) + x\left(e^{-x} - e^{-1}\cos(24\pi x)\right).$$

In this case, we know that the solution will decay in time. The parameters M, h, τ are as before. In Figure 4.9 we plotted the initial solution and in Figure 4.10 we have the numerical solution at four different times. It is clear that the numerical solution is approaching zero for all x. Therefore, the numerical solution of our experiment fully agrees with the theoretical results.

We have experimented with several other initial solutions, and in all cases we noticed the decay of the numerical solution as expected.

4.3 Galerkin's solution of Model II

In the present section questions of the numerical realization of the Galerkin's method approximation (3.132), (3.133) investigated in section 3.6 of Chapter 3 for initial-boundary value problem (3.34)-(3.36) are considered.

For the numerical solution of (3.132), (3.133) we let $\phi_1(x), \ldots, \phi_N(x)$ be a basis for H_h. Therefore, $u^h \in H_h$ can be represented by

$$u^h(x, t) = \sum_{j=1}^{N} u_j(t)\phi_j(x). \tag{4.9}$$

Since (3.132), (3.133) are valid for all $v^h \in H_h$, one can let $v^h = \phi_k$. This yields the following system for the weights $\mathbf{u}(t)$:

$$M\dot{\mathbf{u}} + K(u)\mathbf{u} = \mathbf{F}, \tag{4.10}$$

$$M\mathbf{u}(0) = \mathbf{U}, \tag{4.11}$$

where

$$M_{jk} = \langle \phi_k, \phi_j \rangle, \tag{4.12}$$

$$K(u)_{jk} = \left\langle (1 + S_h(t))\phi'_k, \phi'_j \right\rangle, \tag{4.13}$$

$$\mathbf{F}_j = \langle \phi_j, f \rangle, \qquad \mathbf{U}_j = \langle \phi_j, U_0 \rangle. \tag{4.14}$$

Now we can evaluate $S_h(t)$ as follows

$$\begin{aligned} S_h(t) &= \int_0^t \int_0^1 \left(\sum_{\ell=1}^N u_\ell(\tau)\phi'_\ell(x) \right)^2 dx d\tau \\ &= \int_0^t \int_0^1 \sum_{\ell=1}^N \sum_{m=1}^N u_\ell(\tau) u_m(\tau) \phi'_\ell(x) \phi'_m(x) dx d\tau \\ &= \sum_{\ell=1}^N \sum_{m=1}^N \tilde{K}_{\ell m} \int_0^t u_\ell(\tau) u_m(\tau) d\tau, \end{aligned} \tag{4.15}$$

where

$$\tilde{K}_{\ell m} = \int_0^1 \phi'_\ell(x)\phi'_m(x) dx.$$

The time integral can be approximated by the trapezoidal rule $(0 = t_0 < t_1 < t_2 < \cdots < t_n = t)$ as follows

$$\int_0^t u_\ell(\tau) u_m(\tau) d\tau = \sum_{p=0}^n \tau \zeta_p u_\ell(t_p) u_m(t_p), \tag{4.16}$$

where $\zeta_p = 1/2$ for $p = 0, n$ and $\zeta_p = 1$ for $p = 1, \ldots, n-1$. Combining (4.16) and (4.15) with (4.13), we get

$$\begin{aligned} K(u)_{jk} &= \left\langle \left(1 + \sum_{\ell=1}^{N}\sum_{m=1}^{N} \tilde{K}_{\ell m} \sum_{p=0}^{n} \tau\zeta_p u_\ell(t_p) u_m(t_p)\right) \phi'_k, \phi'_j \right\rangle \\ &= \left(1 + \sum_{\ell=1}^{N}\sum_{m=1}^{N} \tilde{K}_{\ell m} \sum_{p=0}^{n} \tau\zeta_p u_\ell(t_p) u_m(t_p)\right) \tilde{K}_{jk} \\ &= \left(1 + \sum_{p=0}^{n} \tau\zeta_p v(t_p)\right) \tilde{K}_{jk}, \end{aligned} \tag{4.17}$$

where $v(t) = \mathbf{u}^T(t)\tilde{K}\mathbf{u}(t)$. To solve system (4.10), (4.11), we use Taylor's series. Let

$$\mathbf{u}(t+\tau) = \mathbf{u}(t) + \tau\dot{\mathbf{u}}(t) + \frac{1}{2}\tau^2\ddot{\mathbf{u}}(t) + O\left(\tau^3\right). \tag{4.18}$$

Differentiating (4.10) with respect to t, one has

$$M\ddot{\mathbf{u}} + K(u)\dot{\mathbf{u}} + \dot{K}\mathbf{u} = \dot{\mathbf{F}}, \tag{4.19}$$

where

$$\begin{aligned} \dot{K}_{kj} &= \left\langle \dot{S}_h \phi'_j, \phi'_k \right\rangle \\ &= \left\langle \int_0^1 \left(\sum_{\ell=1}^{N} u_\ell(t)\phi'_\ell\right)^2 dx\, \phi'_j, \phi'_k \right\rangle \\ &= \sum_{\ell=1}^{N}\sum_{m=1}^{N} u_\ell(t) u_m(t) \tilde{K}_{\ell m} \tilde{K}_{kj} \\ &= \mathbf{u}^T(t)\tilde{K}\mathbf{u}(t)\tilde{K}_{kj} \end{aligned}$$

$$\dot{K}_{kj} \;=\; v(t)\tilde{K}_{kj}. \tag{4.20}$$

Now multiplying (4.18) by M and using (4.10), (4.19), and (4.20), we have after dropping terms of order higher than two

$$\begin{aligned}
&M\left(\mathbf{u}(t+\tau)-\mathbf{u}(t)\right) = \tau M\dot{\mathbf{u}}(t) + \tfrac{1}{2}\tau^2 M\ddot{\mathbf{u}}(t)\\
&= \tau\left[\mathbf{F}-K(u)\mathbf{u}\right]\\
&+\frac{1}{2}\tau^2\left[\dot{\mathbf{F}}-K(u)\dot{\mathbf{u}}-\dot{K}\mathbf{u}\right]\\
&= \tau\left[\mathbf{F}-K(u)\mathbf{u}\right]\\
&+\frac{1}{2}\tau^2\left[\dot{\mathbf{F}}-K(u)M^{-1}\left(\mathbf{F}-K(u)\mathbf{u}\right)-\dot{K}\mathbf{u}\right]\\
&= \tau\left[\mathbf{F}+\frac{1}{2}\tau\dot{\mathbf{F}}-\frac{1}{2}\tau K(u)M^{-1}\mathbf{F}\right]\\
&-\tau K(u)\left[\mathbf{u}-\frac{1}{2}\tau M^{-1}K(u)\mathbf{u}\right]-\frac{1}{2}\tau^2\dot{K}\mathbf{u}.
\end{aligned} \tag{4.21}$$

If we take $t = t_n$ as in (4.16) and denote $\mathbf{u}^n = \mathbf{u}(t_n)$ then substituting for K and $\dot{K}$ from (4.17) and (4.20), we get

$$\begin{aligned}
&M\left(\mathbf{u}^{n+1}-\mathbf{u}^n\right) = \tau\left[\mathbf{F}^n+\frac{1}{2}\tau\dot{\mathbf{F}}^n\right.\\
&\left.-\frac{1}{2}\tau\left(1+\sum_{p=0}^{n}\tau\zeta_p v(t_p)\right)\tilde{K}M^{-1}\mathbf{F}^n\right]\\
&-\tau\left(1+\sum_{p=0}^{n}\tau\zeta_p v(t_p)\right)\tilde{K}\left[\mathbf{u}^n-\frac{1}{2}\tau M^{-1}\right.\\
&\left.\left(1+\sum_{p=0}^{n}\tau\zeta_p v(t_p)\right)\tilde{K}\mathbf{u}^n\right]-\frac{1}{2}\tau^2 v(t)\tilde{K}\mathbf{u}^n.
\end{aligned} \tag{4.22}$$

Now let us denote

$$\psi^n = 1+\sum_{p=0}^{n}\tau\zeta_p v^p, \tag{4.23}$$

then

$$
\begin{aligned}
M\mathbf{u}^{n+1} &= M\mathbf{u}^n - \tau\psi^n\tilde{K}\left[\mathbf{u}^n - \frac{1}{2}\tau M^{-1}\psi^n\tilde{K}\mathbf{u}^n\right] \\
&- \frac{1}{2}\tau^2 v^n\tilde{K}\mathbf{u}^n + \tau\left[\mathbf{F}^n + \frac{1}{2}\tau\dot{\mathbf{F}}^n\right. \qquad (4.24) \\
&\left. -\frac{1}{2}\tau\psi^n\tilde{K}M^{-1}\mathbf{F}^n\right].
\end{aligned}
$$

In our first numerical experiment we have chosen the right-hand side of (3.53) so that the exact solution is given by

$$U(x,t) = x(1-x)\sin(x+t).$$

In this case the right-hand side is

$$
\begin{aligned}
f(x,t) &= x(1-x)\cos(x+t) - \big(1 + \tfrac{11}{60}t \\
&-\tfrac{1}{8}\cos(t)\sin(t) - \tfrac{1}{8}\cos(1+t)\sin(1+t) \\
&+\tfrac{1}{8}\cos(1)\sin(1)\big)\,(-2\sin(x+t) \\
&+2(1-x)\cos(x+t) - 2x\cos(x+t) \\
&-x(1-x)\sin(x+t))\,.
\end{aligned}
$$

The parameters used are $M = 100$ which dictates $h = 0.01$. In the next two figures we plotted the numerical solution (marked with +) and the exact solution at $t = 0.5$ (Fig. 4.11) and $t = 1.0$ (Fig. 4.12) and it is clear that the two solutions are identical.

We have run the same example with various values of h and measured the error $|||u - u_h|||_1$. The results are given in Table 4.1.

We have experimented with several other initial solutions, and in all cases we noticed the agreement with the exact solution.

Let us note that, clearly, in general, the finite element approximation requires larger storage. In this case, the nonlinear system was solved using Taylor series approximation and one has to solve a banded system at each time step. No iteration is required. On the other hand, in the finite difference case, we have solved the nonlinear system using Newton's method which required iterating with dense matrices. Both methods yield comparable numerical results.

h	error	rate
.2	.0071411	.964499
.04	.0015122	.978861
.02	.00076726	.988986
.01	.00038657	.994375
.005	.00019404	

Table 4.1: The errors for various values of the grid spacing and the approximate rate of convergence

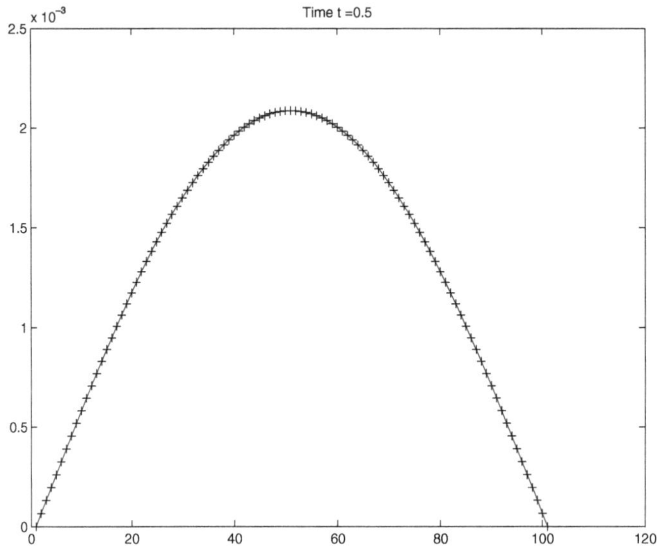

Figure 4.1: The solution at $t = 0.5$. The exact solution is solid line and the numerical solution is marked by $+$.

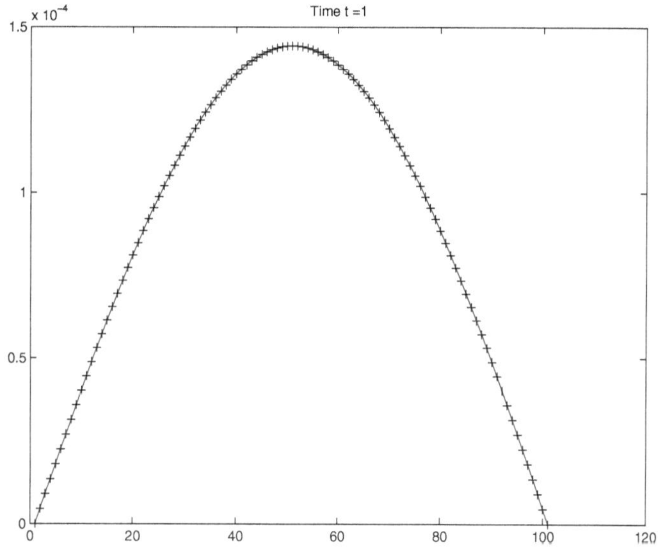

Figure 4.2: The solution at $t = 1.0$. The exact solution is solid line and the numerical solution is marked by $+$.

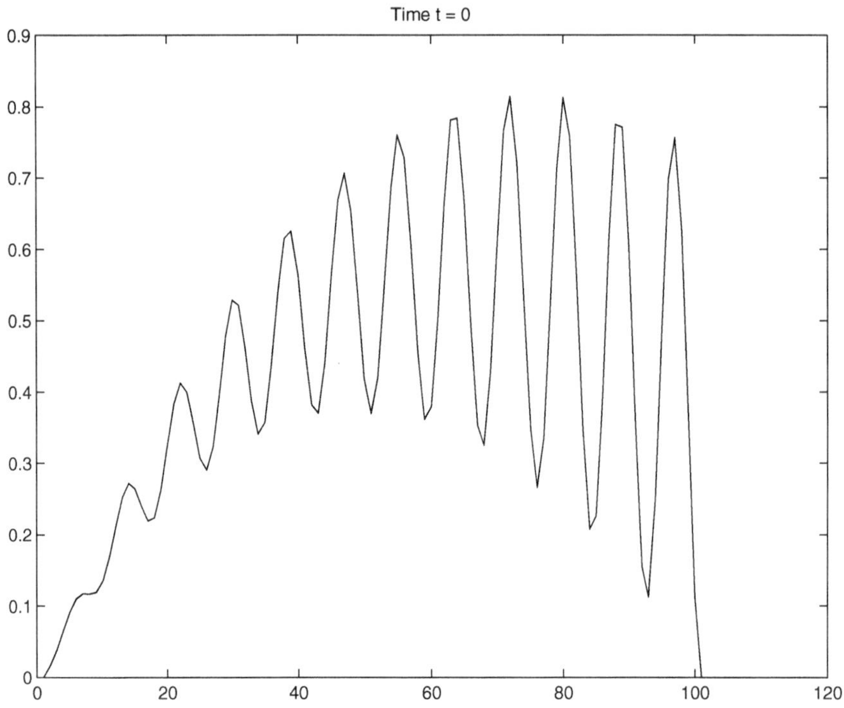

Figure 4.3: The initial data for homogeneous boundary conditions.

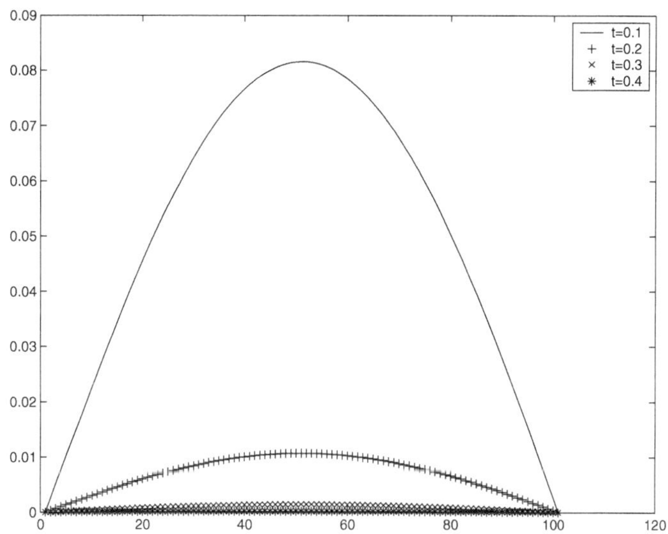

Figure 4.4: The numerical solution at $t = 0.1, 0.2, 0.3, 0.4$ for homogeneous boundary conditions.

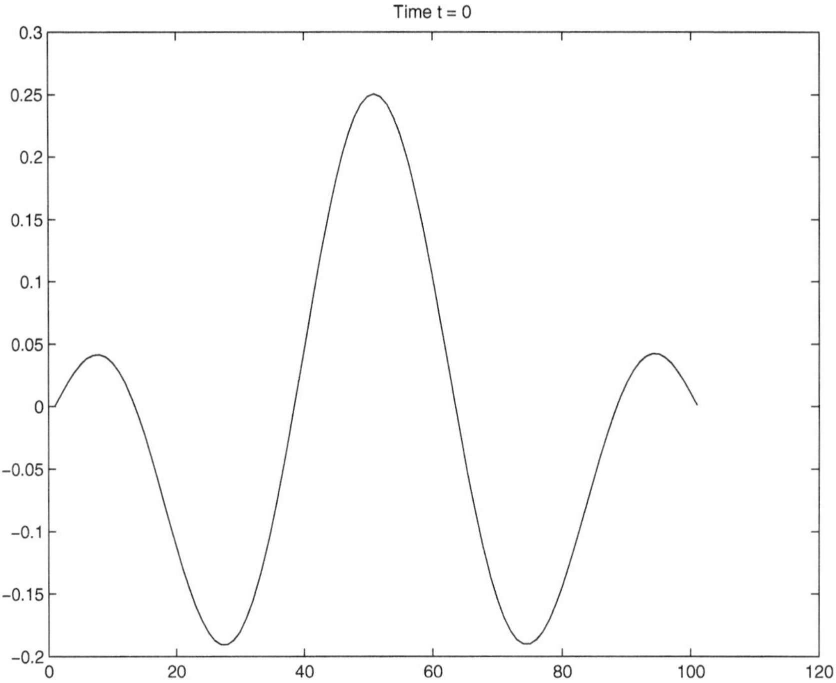

Figure 4.5: The initial data for nonhomogeneous boundary condition on part of lateral boundary.

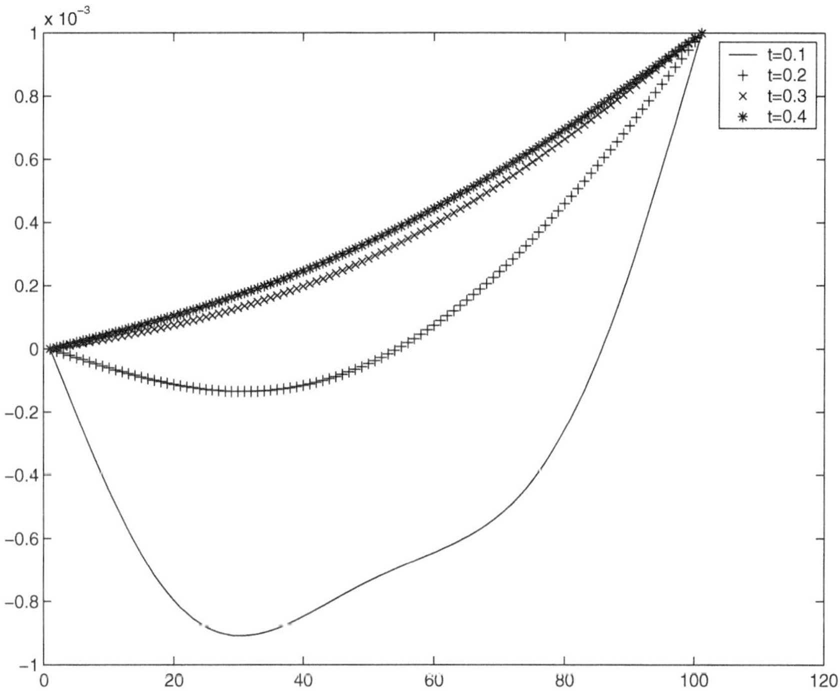

Figure 4.6: The numerical solution at $t = 0.1, 0.2, 0.3, 0.4$ for nonhomogeneous boundary condition on part of lateral boundary.

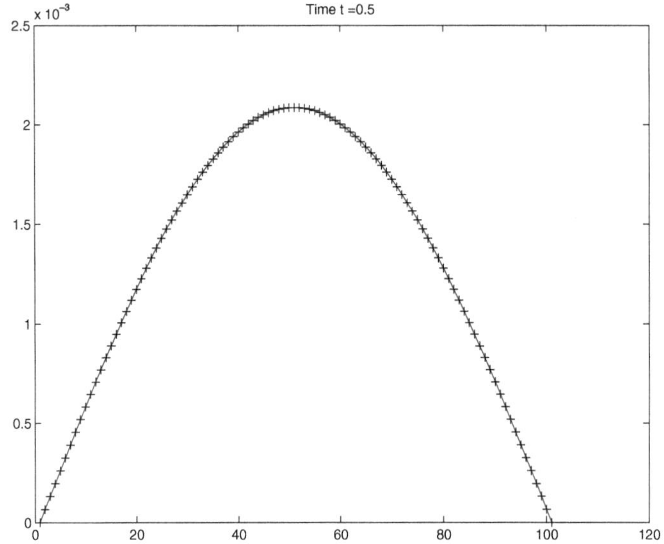

Figure 4.7: The solution at $t = 0.5$. The exact solution is solid line and the numerical solution is marked by +.

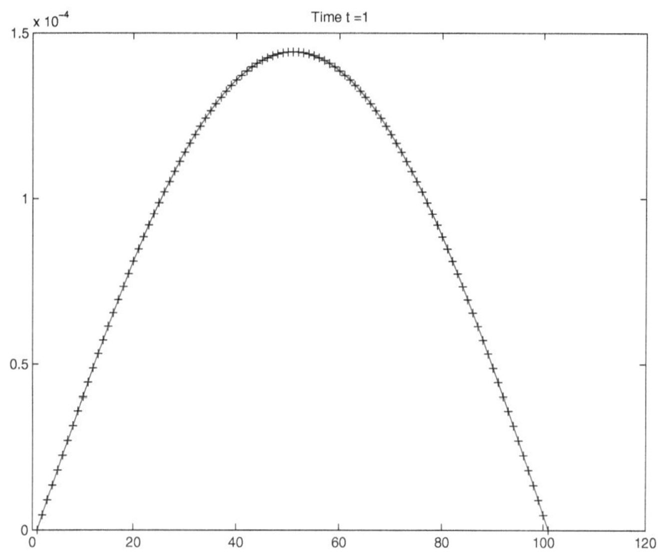

Figure 4.8: The solution at $t = 1.0$. The exact solution is solid line and the numerical solution is marked by $+$.

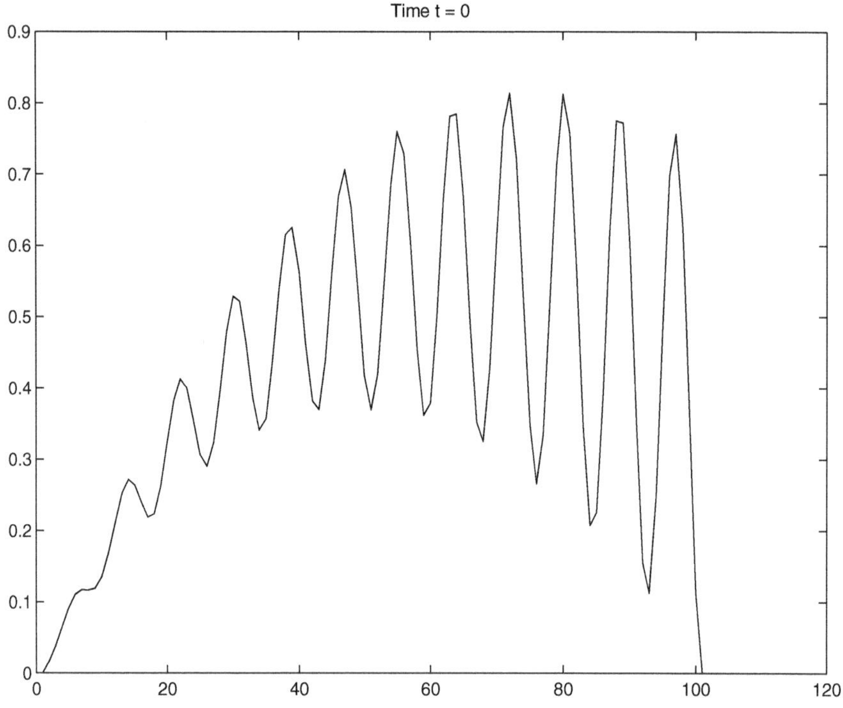

Figure 4.9: The initial solution.

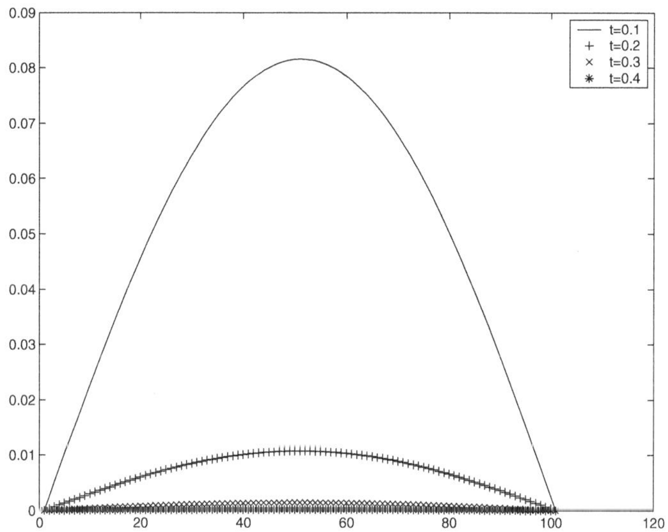

Figure 4.10: The numerical solution at $t = 0.1, 0.2, 0.3, 0.4$.

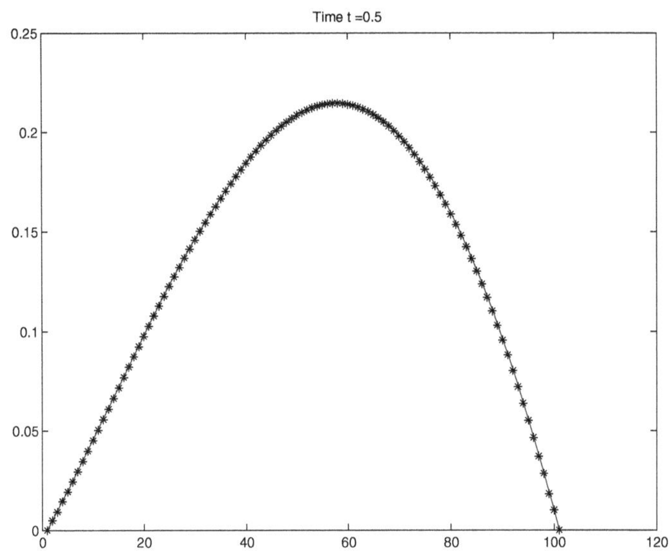

Figure 4.11: The solution at $t = 0.5$. The exact solution is solid line and the numerical solution is marked by +.

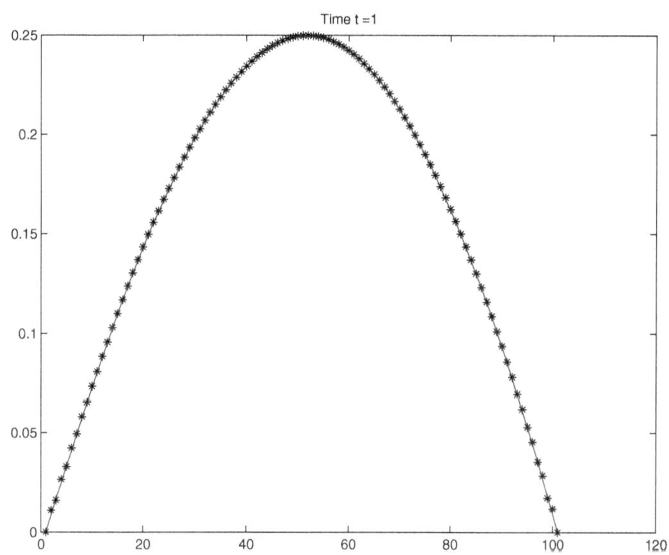

Figure 4.12: The solution at $t = 1.0$. The exact solution is solid line and the numerical solution is marked by $+$.

4.4 Finite difference solution of Model III

In this section we present some of the numerical results obtained when solving (3.78). We have not used homogeneous boundary conditions so that the exact solution will match the initial and boundary conditions.

In our first example, we have the following:

$$a(t) = e^{-t},$$

$$\sigma(p) = p^2,$$

and the exact solution is

$$U(x,t) = e^{-x-t}.$$

Therefore, the initial condition is

$$U_0(x) = e^{-x},$$

and the boundary conditions are:

$$U(0,t) = e^{-t}, \quad U(1,t) = e^{-1-t}.$$

We have found that the forcing term becomes

$$f(x,t) = e^{-x-t} + 2e^{-2x}\left(e^{-t} - e^{-2t}\right).$$

The parameters used are $M = 100$ which dictates $h = 0.01$. In the next two figures we plotted the numerical solution (marked with +) and the exact solution at $t = 0$ (Fig. 4.13) and $t = 0.0062516$ (Fig. 4.14) and it is clear that the two solutions are identical.

In our second example, we have only changed the σ to

$$\sigma(p) = 1 + p^2.$$

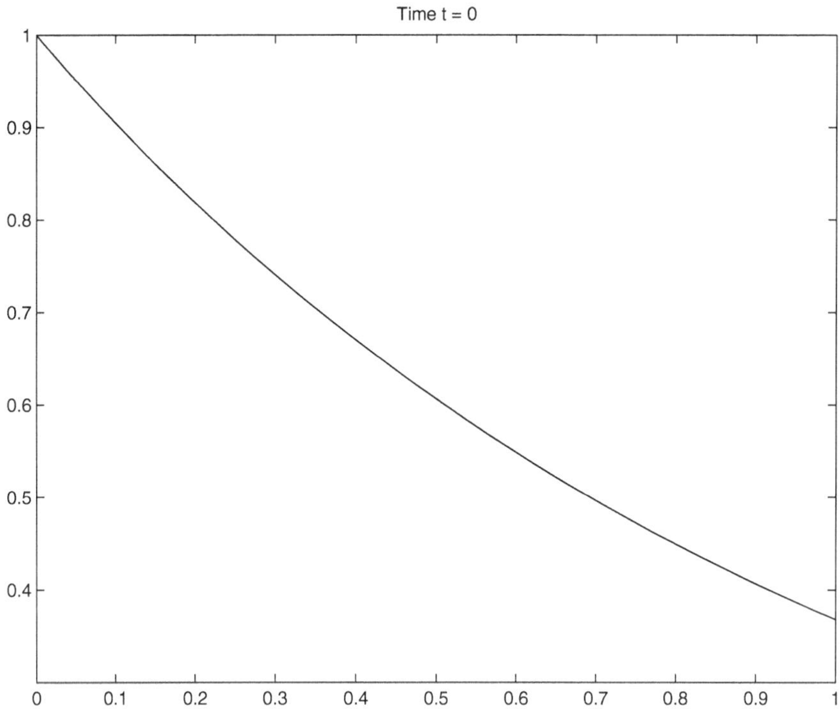

Figure 4.13: The solution at $t = 0$.

All other functions stayed as in the first example. In the next two figures we plotted the numerical solution (marked with +) and the exact solution at $t = 0$ (Fig. 4.15) and $t = 0.0062516$ (Fig. 4.16) and it is clear that the two solutions are identical.

In our first example, we have the following:

$$a(t) = e^{-2t},$$

$$\sigma(p) = p^2,$$

and the exact solution is

$$U(x, t) = \sin(x + t).$$

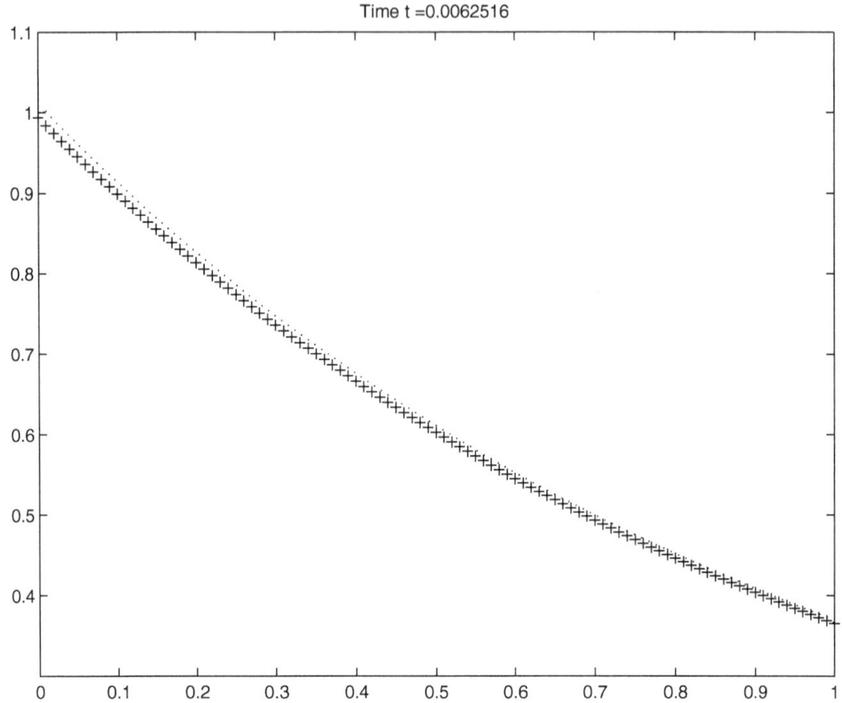

Figure 4.14: The solution at $t = 0.0062516$. The exact solution is solid line and the numerical solution is marked by $+$.

Therefore, the initial condition is

$$U_0(x) = \sin(x)$$

and the boundary conditions are:

$$U(0,t) = \sin(t), \quad U(1,t) = \sin(1+t).$$

We have found that the forcing term becomes

$$f(x,t) = \cos(x+t) + \frac{1}{4}\left(\sin(2(x+t)) - \cos(2(x+t))\right)$$
$$-e^{-2t}\left(\sin(2x) - \cos(2x)\right).$$

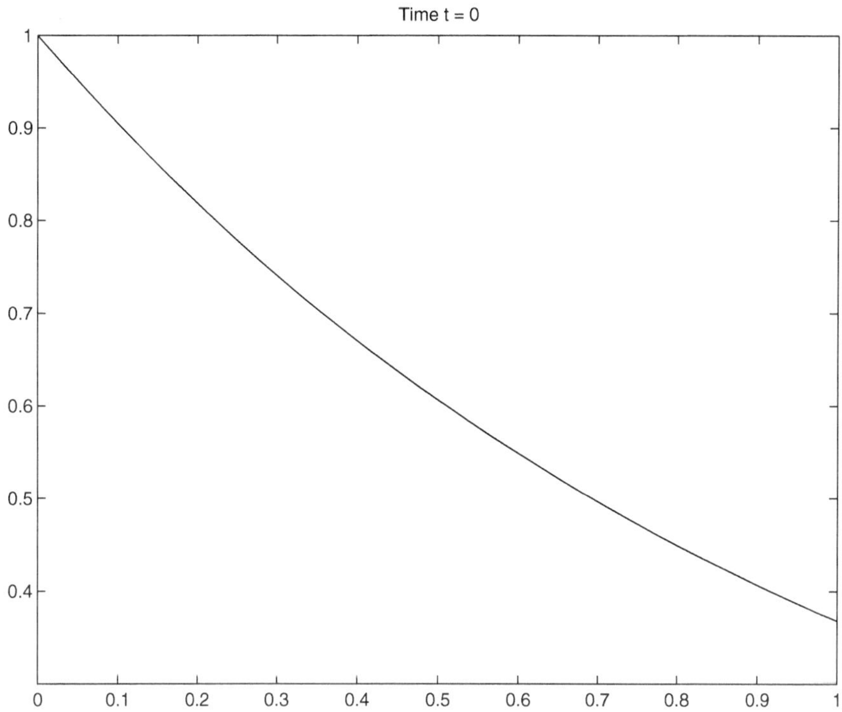

Figure 4.15: The solution at $t = 0$.

The parameters used are $M = 100$ which dictates $h = 0.01$. In the next two figures we plotted the numerical solution (marked with +) and the exact solution at $t = 0$ (Fig. 4.17) and $t = 0.0062516$ (Fig. 4.18) and it is clear that the two solutions are identical.

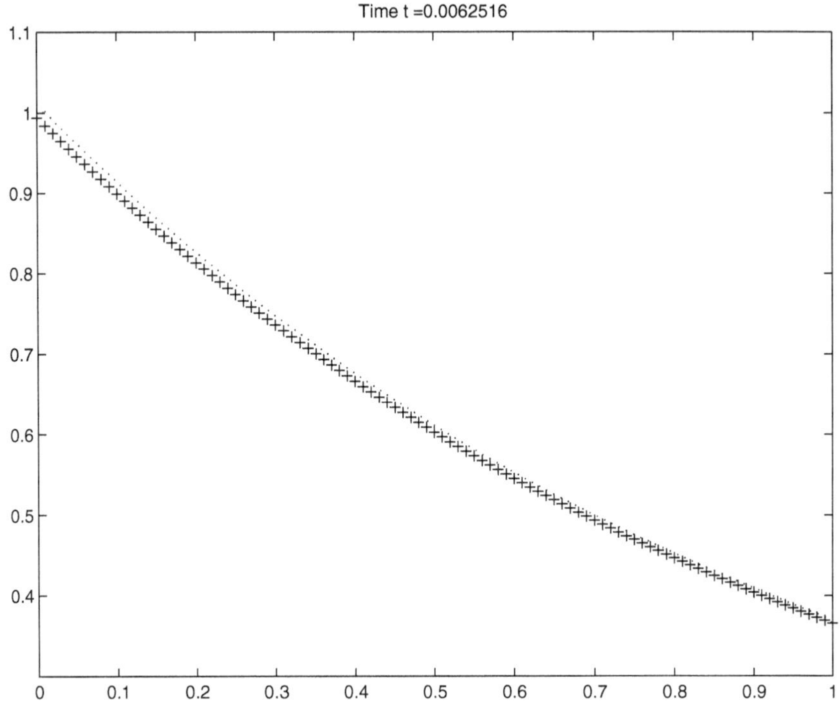

Figure 4.16: The solution at $t = 0.0062516$. The exact solution is solid line and the numerical solution is marked by $+$.

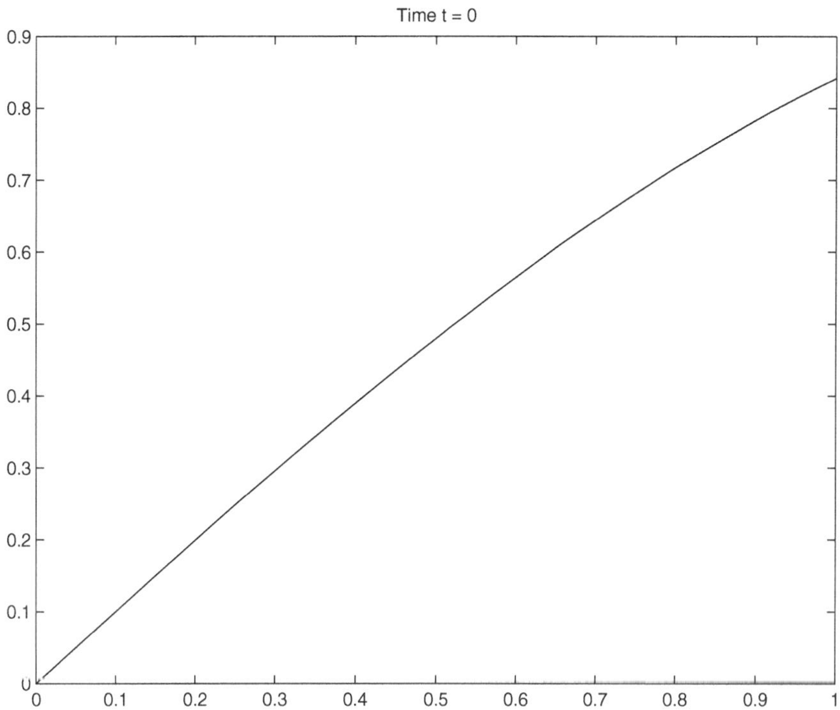

Figure 4.17: The solution at $t = 0$.

4.5 Comments and bibliographical notes

The last step before resolving discrete analogues on the computer is to construct and use the algorithms of solution of the systems of numerical equations. There are many scientific papers in this direction also, see, for example, [23], [68], [167], [183], [193], [211], [301], [371], [400], [406], [407], [411], [435],

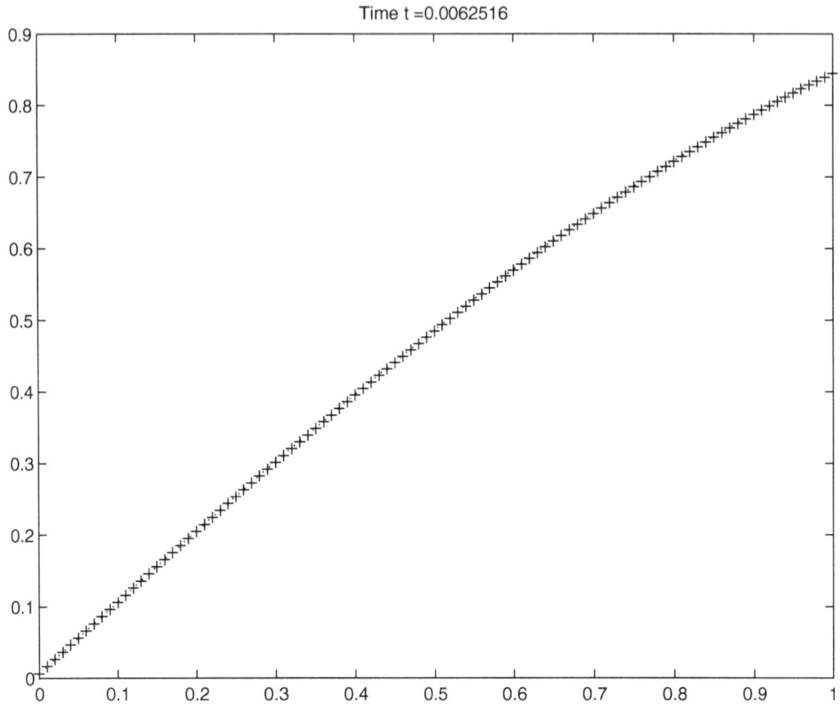

Figure 4.18: The solution at $t = 0.0062516$. The exact solution is solid line and the numerical solution is marked by $+$.

[440], [446], and references therein.

Computational electromagnetism and heat processes refers to the modern concept of computer-aided analysis. There are many works in this direction. For example, the book [68] is uniquely written to enable the reader to successfully perform important simulation techniques and to design efficient computer software for electromagnetic device analysis. Numerous illustrations, solved exercises, original ideas, and an extensive and up-to-date bibliography make it a valuable reference for both experts and beginners in the field. A researcher and practitioner will find in it information rarely available in other sources, such as on symmetry, bilateral error bounds by complementarity, edge and face elements, treatment of infinite domains, etc. As a self-contained text, it presents an extensive coverage of the most important concepts from Maxwell's equations to computer-solvable algebraic systems - for static, quasi-static, and harmonic high-frequency problems.

Numerical solutions of the considered integro-differential models have their specifics and we have noted those in the previous chapter.

One must note that many scientific works are devoted to the algorithms of numerical realization of methods of approximate solution for Maxwell's and like it systems of nonlinear partial differential equations (see, for example, [54], [68], [72], [73], [139], [145], [148], [153], [173], [195], [197], [205], [210], [212], [276], [314], [315], [316], [317], [336], [341], [369], [442], [472], [480], [493], and [504]). In most of these works results of numerical experiments are given with graphical illustrations and their analysis.

Now we give more detailed bibliographic references and short comments for the separate paragraphs of Chapter 4.

4.1. Numerical solution of Model I

Investigation and numerical solution of the semi-discrete and discrete schemes for (2.22) type integro-differential models are started in the work [139]. There are many numerical experiments performed here. Analysis of the theoretical mathematical researches and physical problems too is given in this work.

Many scientific works are dedicated to numerical solution of (2.22) type equations [139], [145], [220], [227], [239], [243], [244], [276], [280], [318], [333], and [334]. Numerical algorithms based on the above-mentioned type schemes and numerical results with graphical illustrations and their analysis are given in [139], [243], [244], and [280].

Numerical methods for solving one-dimensional integro-differential systems of type (2.22) with two-component magnetic field are studied in the works [274] and [276]. Let us note that numerical results and graphical illustrations are also given in [255].

In section 4.1 we gave some graphical illustrations which are received by finite difference method. The analogical results can be obtained by Galerkin's method.

4.2. Numerical solution of Model II

Investigation and numerical resolution of the semi-discrete and discrete schemes for (2.23) type integro-differential models were initiated in [145].

Many scientific works are also dedicated to numerical solution of (2.23) type equations: [31], [32], [96], [145], [239], [243], [256], [423], [424], and [501]. Various numerical results and corresponding graphical figures are given in [243] and [256].

Let us note that such investigations for one-dimensional integro-differential (2.23) type averaged systems with two-component magnetic field are also studied in the works [33], [223], [241], [255], [257], and [276]. Illustrations of the appropriate numerical experiments are given in [257].

For the (2.23) type averaged integro-differential models numerical results obtained by Galerkin's finite element algorithm are described in [258], [423], [424], and [501].

Let us note that such numerical results and figures for one-dimensional integro-differential (2.23) type averaged systems with two-component magnetic field are also given in [261].

4.3. Numerical solution of Model III

In [378] difference schemes and corresponding algorithm to obtain a numerical solution to the integro-differential equation describing heat flow in materials with memory are studied. In [376] Galerkin's method is also used for numerical solution of this model. Analysis of the corresponding numerical results and graphical illustrations is given in these works as well. The issues concerning with the numerical solution of Model III by various methods are given in many other works as well (see, for example, [20], [80], [125], [401], and reference therein).

As we have already said, many numerical experiments are done using the constructed and investigated algorithms. Let us note that for numerical solution of the given discrete analogues Newton's iterative method [406] and sometimes its modifications are applied (see [139], [145], and [148]).

Bibliography

[1] Abassy, T.A., El-Tawil, M.A., Saleh, H.K., The solution of KdV and mKdV equations using adomian pade approximation, Int. J. Nonlinear Sci. Numer. Simul., 5, (2004), 327–340.

[2] Abbasbandy, S., A numerical solution of Blasius equation by Adomian's decomposition method and comparison with homotopy perturbation method, Chaos Solitons Fractals, 31, (2007), 257–260.

[3] Abbasbandy, S., Application of He's homotopy perturbation method to functional integral equations, Chaos Solitons Fractals, 31, (2007), 1243–1247.

[4] Abuladze, I.O., Gordeziani, D.G, Dzhangveladze, T.A., Korshiya T.K., Discrete models of a nonlinear magnetic-field-scattering problem with thermal conductivity, Differ. Uravn., 22 (7), (1986), 1119–1129 (in Russian). English translation: Diff. Equ., 22 (7), (1986), 769–777.

[5] Abuladze, I.O., Gordeziani, D.G, Dzhangveladze, T.A., Korshiya T.K., On the numerical modeling of a nonlinear problem of the diffusion of a magnetic field with regard to heat conductivity, Proc. I. Vekua Inst. Appl. Math. (Tbiliss. Gos. Univ. Inst. Prikl. Math. Trudy), 18, (1986), 48–67 (in Russian).

[6] Acquistapace, P., Terreni, B., Existence and sharp regularity results for linear parabolic non-autonomous integro-differential equations, Israel J. Mathematics, 53 (3), (1986), 257–303.

[7] Adams, R.A., Sobolev Spaces, Academic Press, New York, 1975.

[8] Adomian, G., Frontier Problem of Physics: The Decomposition Method, Kluwer Academic Publish, Boston, 1994.

[9] D'Ancona, P., Christie, I., Sanz-Serna, J.M., A Galerkin method for a nonlinear integro-differential wave system. Comput. Methods Appl. Mech. Engrg., 44, (1984), 229–237.

[10] Anselone, P.M., Nonlinear Integral Equations, University Wisconsin Press, Madison, 1964.

[11] Agarwal, R., Bohner, M., Domoshnitsky, A., Goltser, Y., Floquet theory and stability of nonlinear integro-differential equations, Acta Math. Hungar. 109 (4), (2005), 305–330.

[12] Aguilar, M., Brunner, H., Collocation methods for second-order Volterra integro-differential equations, Appl. Numer. Math., 4, (1988), 455–470.

[13] Aizicovici, S., Time-dependent Volterra integrodifferential equations, J. Integral Equations, 10, (1985), Suppl., 45–60.

[14] Aizicovici, S., Hannsgen, K.B., Local existence for abstract semilinear Volterra integro-differential equations, J. Integral Equations, 5, (1993), 299–313.

[15] Akrivis, G., Larsson, S., Linearly implicit finite element methods for the time-dependent Joule heating problem, BIT, 45, (2005), 429–442.

[16] Alawneh, A., Al-Khaled, K., Al-Towaiq, M., Reliable algorithms for solving integro-differential equations with applications, Int. J. Comput. Math., 87 (7), (2010), 1538–1554.

[17] Alduncin, G., Oden, J.T., Analysis of Galerkin approximations of a class of pseudomonotone diffusion problem, SIAM J. Math. Anal., 12, (1981), 917–930.

[18] Allegretto, W., Lin, Y., Zhou, A., Long-time stability of finite element approximations for parabolic equations with memory, Numer. Methods Partial Differential Equations, 15 (3), (1999), 333–354.

[19] Allegretto, W., Ningning Yan, N., A posteriori error analysis for FEM of thermistor problems, Int. J. Numer. Anal. Model., 3 (4), (2006), 413–436.

[20] Alquran, M., Al-Khaled. K., Approximate solutions to nonlinear partial integro-differential equations with applications in heat flow, Jordan J. Math. Stat., 3 (2), (2010), 93–116.

[21] Amadori, A., Nonlinear integro-differential evolution problems arising in option pricing: A viscosity solutions approach, Differ. Integral Equ., 16 (7), (2003), 787–811.

[22] Amadori, A.L., Karlsen, K.H., La Chioma, C., Nonlinear degenerate integro-partial differential evolution equations related to geometric Levy processes and applications to backward stochastic differential equations, Stochastics and Stochastic Reports, 76, (2004), 147–177.

[23] Ames, W.F., Numerical Methods for Partial Differential Equations, Academic Press, New York, 1977.

[24] D'Ancona P., Spagnolo S., Nonlinear perturbations of the Kirchhoff equation, Comm. Pure Appl. Math., 47, (1994), 1005–1029.

[25] Antontsev, S.N., Kazhikhov, A.V., Monakhov, V.N., Boundary Value Problems of Inhomogeneous Fluid Mechanics, Studies in Mathematics and Its Applications, North-Holland, 1990.

[26] Aptsiauri, M. M., Three-level difference schemes for one nonlinear system of partial differential equations, Rep. Enlarged Sess. Semin. I. Vekua Appl. Math., 10, (1995), 1–3.

[27] Aptsiauri, M. M., Finite difference schemes for one system of nonlinear partial differential equations, Proc. I. Vekua Inst. Appl. Math., 50-51, (2000-2001), 1–4.

[28] Aptsiauri, M., On one averaged integro-differential model, Rep. Enlarged Sess. Semin. I. Vekua Appl. Math., 25, (2011), 10–13.

[29] Aptsiauri, M., Asymptotic behavior of the solution of one nonlinear integro-differential model with source term, Rep. Enlarged Sess. Semin. I. Vekua Appl. Math., 26, (2012), 1–4.

[30] Aptsiauri, M., Gagoshidze, M., On one nonlinear averaged integro-differential system with source terms, Rep. Enlarged Sess. Semin. I. Vekua Appl. Math., 28, (2014), 5–8.

[31] Aptsiauri, M., Jangveladze, T., Kiguradze, Z., Large time behavior of solutions and numerical approximation of nonlinear integro-differential equation associated with the penetration of a magnetic field into a substance, Appl. Math. Inform. Mech., 13 (2), (2008), 3–17.

[32] Aptsiauri, M., Jangveladze, T., Kiguradze, Z., On the stabilization of solution as $t \to \infty$ and convergence of the corresponding finite difference scheme for one nonlinear integro-differential equation, Rep. Enlarged Sess. Semin. I. Vekua Appl. Math., 22, (2008), 15–19.

[33] Aptsiauri, M., Jangveladze, T., Kiguradze, Z., On asymptotic behavior of solution of one nonlinear one-dimensional integro-differential analogue of Maxwell's system, Rep. Enlarged Sess. Semin. I. Vekua Appl. Math., 23, (2009), 5–10.

[34] Aptsiauri, M.M., Jangveladze, T.A., Kiguradze, Z.V., Asymptotic behavior of the solution of a system of nonlinear integro-differential equations, Differ. Uravn., 48, (2012), 70–78 (in Russian). English translation: Differ. Equ., 48 (2012), 72–80.

[35] Aptsiauri, M., Kiguradze, Z., On one nonlinear integro-differential equation, Rep. Enlarged Sess. Semin. I. Vekua Appl. Math., 24, (2010), 9–12.

[36] Arosio, A., Averaged evolution equations. The Kirchhoff string and its treatment in scales of Banach spaces, Second workshop on functional-analytic methods in complex analysis and applications to partial differential equations, Trieste, 1993, 220–254, World Sci. Publ., River Edge, NJ, 1995.

[37] Arosio, A., Panizzi, S., On the well-posedness of the Kirchhoff string, Trans. Amer. Math. Soc., 348, (1996), 305–330.

[38] Assous, F., Degond, P., Heintze, E., Raviart, P.A., Segré, J., On a finite-element method for solving the three-dimensional Maxwell equations, J. Comput. Phys., 109 (2), (1993), 222–237.

[39] Assous, F., Ciarlet Jr, P., Segré, J., Numerical solution to the time-dependent Maxwell equations in two-dimensional singular domains: the singular complement method, J. Comput. Phys., 161 (1), (2000), 218–249.

[40] Attigui, F., Contróle et Stabilisation de Cábles Non Linéaires, Repport de stage scientifique, Laboratoire des Materiaux et des Structures du Genie Civil, Universite Paris XIII, France, 1999.

[41] Atkinson, K., A Survey of Numerical Methods for the Solution of Fredholm Integral Equations of the Second Kind, Society for Industrial and Applied mathematics, Philadelphia, 1976.

[42] Atkinson, K.E., The Numerical Solution of Integral Equations of the Second Kind, Cambridge University Press, Cambridge (1997).

[43] Babuska, I., Aziz, A.K., Survey lectures on the mathematical foundations of the finite element method, The Mathematical Foundations of the Finite Element Method with Applications to Partial Differential Equations (Proc. Sympos., Univ. Maryland, Baltimore, Md., 1972), Academic Press, New York, 1972.

[44] Bahuguna, D., Integrodifferential equations with analytic semigroups, J. Applied Math. Stochastic Analysis, 16, (2003), 177–189.

[45] Bahuguna, D., Raghavendra, V., Rothe's method to parabolic integrodifferential equations via abstract integrodifferential equations, Appl. Anal., 33, (1989), 153–167.

[46] Bahuguna, D., Shukla, R., Method of semidiscretization in time for quasilinear integrodifferential equations, Int. J. Math. Math. Sci., 2004, (2004), 469–478.

[47] Bahuguna, D., Srivastava, S.K., Approximation of solutions to evolution integrodifferential equations, J. Applied Math. Stochastic Analysis, 9, (1996), 315–322.

[48] Baker, C.T.H., The Numerical Treatment of Integral Equations: Monographs on Numerical Analysis, Clarendon Press, Oxford, 1977.

[49] Bai, Y., Backward solutions to nonlinear integro-differential systems, Cent. Eur. J. Math., 8, (2010), 807–815.

[50] Bai, Y., Zhang, P. On a class of Volterra nonlinear equations of parabolic type, Appl. Math. Comput., 216, (2010), 236–240.

[51] Barbu, V., Integro-differential equations in Hilbert spaces, Anal. St. Univ. "Al. I. Cuza", 19, (1973), 365–383.

[52] Barbu, V., On a nonlinear Volterra equation on a Hilbert space, SIAM J. Math. Anal., 8, (1977), 346–355.

[53] Barbu, V., Iannelli, M., The controllability of the heat equation with memory, Differential Integral Equations, 13, (2001), 1393–1412.

[54] Baxevanis, Th., Katsaounis, Th., Tzavaras, A.E., Adaptive finite element computations of shear band formation, Math. Models Methods Appl. Sci., 20, (2010), 423–448.

[55] Bernstein, S., Sur une classe d'équations fonctionnelles aux dérivées partielles, Izv. Akad. Nauk SSSR, Ser. Mat., 4, (1940), 17–26 (in Russian).

[56] Besenyei, A., On some systems containing a parabolic PDE and a first order ODE, Math. Bohem., 135, (2010), 133–141.

[57] Bhattacharya, S., Mandal, B.N., Numerical solution of a singular integro-differential equation, Appl. Math. Comput., 195, (2008), 346–350.

[58] Bhrawy, A.H., Tohidi, E., Soleymani, F., A new Bernoulli matrix method for solving high-order linear and nonlinear Fredholm integro-differential equations with piecewise intervals, Appl. Math. Comput., 219, (2012), 482–497.

[59] Bian, B., Guan, P., Convexity preserving for fully nonlinear parabolic integro-differential equations, Methods Appl. Anal., 15, (2008), 39–52.

[60] Bień, M., Existence of global weak solutions for a class of quasilinear equations describing Joule's heating, Math. Methods Appl. Sci., 23, (1998), 1275–1291.

[61] Bilbao, S., Modal-type synthesis techniques for nonlinear stringe with an energy conservation property, Proc. Int. Conference on Digital Audio Effects, Naples, Italy, (2004), 119–124.

[62] Birman, M.S., Solomyak, M.Z., L_2-theory of the Maxwell operator in arbitrary domains, Russian Math. Surveys, 42 (6), (1987), 75–96.

[63] Biswas, I.H., Jakobsen, E.R., Karlsen, K.H., Difference-quadrature schemes for nonlinear degenerate parabolic integro-PDE, SIAM J. Numer. Anal., 48, (2010), 1110–1135.

[64] Bitsadze, A.V., On some integral equations of the first kind, Dokl. Akad. Nauk SSSR, 286, (1986), 1292–1295 (in Russian). English translation: Sov. Math. Dokl., 33, (1986), 270–272.

[65] Böhm, M, Taboada, M., Global existence and regularity of solutions of the nonlinear string equation, IMA Preprint Series No. 1078, 1992.

[66] Boltzmann, L., Weitere Studien über das Warmegleichgewicht unter Gasmolekülen, Sitzungsberichte Akademie der Wissenschaften, 66, (1872), 275–370. (included in Wissenschaftliche Abhandlungen, 1, (1909), 316–402).

[67] Bonfanti, G., Luterotti., F., Regularity and convergence results for a phase-field model with memory, Math. Methods Appl. Sci., 21, (1998), 1085–1105.

[68] Bossavit, A., Computational Electromagnetism, Variational Formulations, Complementarity, Edge Elements. Academic Press Series in Electromagnetism, Academic Press Inc., San Diego, CA, 1998.

[69] Bouziani, A., Mechri, R., The Rothe's method to a parabolic integrodifferential equation with a nonclassical boundary conditions, Int. J. Stoch. Anal., 2010, (2010), 1–17.

[70] Brandon, D., Global existence and asymptotic stability for a nonlinear integrodifferential equation modeling heat flow, SIAM J. Math. Anal., 22, (1991), 72–106.

[71] Brenner, S.C., Scott, L.R., The Mathematical Theory of Finite Element Methods, Third Edition, Springer-Verlag, New York, 2008.

[72] Brezzi, F., Lipnikov, K., Shashkov, M., Convergence of the mimetic finite difference method for diffusion problems on polyhedral meshes, SIAM J. Numer. Anal., 43, (2005), 1872–1896.

[73] Brezzi, F., Lipnikov, K., Shashkov, M., Simoncini, V., A new discretization methodology for diffusion problems on generalized polyhedral meshes, Comput. Methods Appl. Mech. Engrg., 196, (2007), 3682–3692.

[74] Briani, M., La Chioma, C., Natalini, R., Convergence of numerical schemes for viscosity solutions to integro-differential degenerate parabolic problems arising in financial theory, Numer. Math., 98, (2004), 607–646.

[75] Browder, F.E., Nonlinear equations of evolution, Ann. of Math., 80, (1964), 485–523.

[76] Browder, F.E., Strongly nonlinear parabolic value problems, Amer. J. Math., 86, (1964), 339–357.

[77] Browder, F.E., Multi-valued monotone nonlinear mappings and duality mappings in Banach spaces, Trans. Amer. Math. Soc., 118, (1965), 338–351.

[78] Brunner, H., A survey of recent advances in the numerical treatment of Volterra integral and integro-differential equations, J. Comput. Appl. Math., 8, (1982), 213–229.

[79] Brunner, H., van der Houwen, P.J., The Numerical Solution of Volterra Equations, CWI Monographs, Vol. 3, North-Holland, Amsterdam, 1986.

[80] Brunner, H., Collocation Methods for Volterra Integral and Related Functional Differential Equations, Cambridge University Press, Cambridge, U.K., 2004.

[81] Brunner, H., Schötzau, D., hp-discontinuous Galerkin timestepping for Volterra integrodifferential equations, SIAM J. Numer. Anal., 44, (2006), 224–245.

[82] Budak, B.M., Pavlov, A.R. Difference methods of solution of boundary value problems for integro-differential equations, Proc. Computing Center of Moscow State University (B.M.Budak, ed.) 1970, 1–82 (in Russian).

[83] Budak, B.M., Pavlov, A.R., A difference method of solving boundary value problems for a quasilinear integro-differential equation of parabolic type, Dokl. Akad. Nauk SSSR, 209, (1973), 1017–1020 (in Russian). English translation: Soviet Math. Dokl., 14, (1973), 565–569.

[84] Cafarelli, L., Silvestre, L. Regularity theory for fully nonlinear integro-differential equations, Comm. Pure Appl. Math., 62, (2009), 597–638.

[85] Camino, P., A numerical method for a partial differential equation with memory, Proc. 9th CEDYA. Secretariado de Publicaciones, Universidad de Valladolid, Valladolid, (1988), 107–112.

[86] Cannon, J.R., Lin, Y., Nonclassical H^1 projection and Galerkin methods for nonlinear parabolic integro-differential equations, Calcolo, 25, (1988), 187–201.

[87] Cannon, J.R., Lin, Y., Smooth solutions for an integro-differential equation of parabolic type, Differential Integral Equations, 2 (1989), 111–121.

[88] Cannon, J.R., Lin, Y., A priori L_2 error estimates for finite-element methods for nonlinear diffusion equations with memory, SIAM J. Numer. Anal., 27, (1990), 595–607.

[89] Capobianco, M.R., Criscuolo, G., Junghanns, P., Luther, U., Uniform convergence of the collocation method for Prandtl's integro-differential equation, ANZIAM J., 42, (2000), 151–168.

[90] Caraus, I., Mastorakis, N.E., The stability of collocation methods for approximate solution of singular integro-differential equations, WSEAS Transaction on Math., 7 (4), (2008), 121–129.

[91] Castro, L.P., Ramos, A., Hyers-Ulam-Rassias stability for a class of nonlinear Volterra integral equations, Banach J. Math. Anal., 3 (1), (2009), 36–43.

[92] Cessenat, M., Mathematical Methods in Electromagnetism Linear Theory and Applications, Series on Advances in Mathematics for Applied Sciences, Volume 41, World Scientific Publishers, Singapore, 1996.

[93] Chadam, J.M., Yin, H.M., An iteration procedure for a class of integrodifferential equations of parabolic type, J. Integral Equations Appl., 2, (1990), 31–47.

[94] Charalambakis, N., Adiabatic shearing flow caused by time-dependent inertial force, Quart. Appl. Math., 42, (1984), 275–280.

[95] Charalambakis, N., Murat, F., Approximation by finite elements, existence and uniqueness for a model of stratified thermoviscoplastic materials, Ric. Mat., 55, (2006), 171–218.

[96] Chen, F., Crank-Nicolson fully discrete-Galerkin mixed finite element approximation of one nonlinear integrodifferential model, Abstract Appl. Anal., Hindawi Publishing Corporation, 2014, (2014), 534902, 8 pages.

[97] Chen, C., Shih, T., Finite Element Methods for Integrodifferential Equations, Series on Appl. Math. Vol. 9, World Scientific, 1998.

[98] Chen, C., Thomée, V., Wahlbin, L.B., Finite element approximation of a parabolic integro-differential equation with a weakly singular kernel, Math. Comp., 58, (1992), 587–602.

[99] Chen, X., Friedman, A., The thermistor problem for conductivity which vanishes at large temperature, Quart. Appl. Math., 51 (1), (1993), 101–115.

[100] Chetverikov, V.N., Kudryavtsev, A.G., Modeling integro-differential equations and a method for computing their symmetries and conservation laws, Amer. Math. Soc. Transl., 167, (1995), 1–22.

[101] Chew, W.C., Weddon, W.H., A 3-D perfectly matched medium from modified Maxwell's equations with stretched coordinates, Micro. Opt. Tech. Lett., 7, (1994), 599–604.

[102] Christie, I., Sanz-Serna, J.M., A Galerkin method for a nonlinear integro-differential wave system, Comput. Methods Appl. Mech. Engrg. 44, (1984), 229–237.

[103] Chrysafinos, K., Approximations of parabolic integro-differential equations using wavelet-Galerkin compression techniques, BIT, 47, (2007), 487–505.

[104] Chrysafinos, K., Walkington, N.J., Error estimates for the discontinuous Galerkin method for parabolic equations, SIAM J. Numer. Anal., 44, (2006), 349–366.

[105] Cimatti, G., Remark on existence and uniqueness for the thermistor problem under mixedboundary conditions, Quart. Appl. Math., 47 (1), (1989), 117–121.

[106] Cimatti, G., Existence of weak solutions for the nonstationary problem of the Joule heating of a conductor, Ann. Mat. Pura Appl., 162, (1992), 33–42.

[107] Clément, P., Nohel, J., Asymptotic behavior of solutions of nonlinear Volterra equations with completely positive kernels, SIAM J. Math. Anal., 12, (1981), 514–535.

[108] Clément, P., Prüss, J., Global existence for a semilinear parabolic Volterra equation, Math. Z., 209, (1992), 17–26.

[109] Coleman, B.D., Gurtin, M.E., Waves in materials with memory II on the growth and decay of one-dimensional acceleration waves, Arch. Ration. Mech. Anal., 19, (1965), 239–265.

[110] Colli, P., Grasselli, M., Hyperbolic phase change problems in heat conduction with memory, Proc. Roy. Soc. Edinburgh Sect. A, 123, (1993), 571–592.

[111] Colombo, F., Quasilinear systems with memory kernels, J. Math. Anal. Appl., 311, (2005), 40–68.

[112] Cont, R., Voltchkova, E., A finite difference scheme for option pricing in jump diffusion and exponential Levy models, SIAM J. Numer. Anal., 43, (2005), 1596–1626.

[113] Cont, R., Voltchkova, E., Integro-differential equations for option prices in exponential Lévy models, Finance Stoch., 9, (2005), 229–325.

[114] Corduneanu, C., Integral Equations and Applications, Cambridge University Press, Cambridge, 1991.

[115] Costabel, M., A coercive bilinear form for Maxwell's equations, J. Math. Anal. Appl.,157 (2), (1991), 527–541.

[116] Crandall, M.G., Londen, S.O., Nohel, J.A., An abstract nonlinear Volterra integrodifferential equation, J. Math. Anal. Appl., 64, (1978), 701–735.

[117] Cuminato, J.A., Fitt, A.D., McKee, S., A review of linear and nonlinear Cauchy singular integral and integro-differential

equations arising in mechanics, J. Integral Equations Appl., 19 (2), (2007), 163–207.

[118] Da, X., On the discretization in time for a parabolic integrodifferential equation with a weakly singular kernel. I: Smooth initial data, Appl. Math. Comput., 58, (1993), 1–27.

[119] Dafermos, C.M., An abstract Volterra equation with application to linear viscoelasticity, J. Differential Equations, 7, (1970), 554–569.

[120] Dafermos, C.M., Stabilizing effects of dissipation, Equadiff 82, Lecture Notes Math., 1017, (1983), 140–147.

[121] Dafermos, C.M., Hsiao, L., Adiabatic shearing of incompressible fluids with temperature-dependent viscosity, Quart. Appl. Math., 41, (1983), 45–58.

[122] Dafermos, C.M., Nohel, J.A., Energy methods for nonlinear hyperbolic Volterra integrodifferential equations, Comm. Partial Differential Equations, 4, (1979), 219–278.

[123] Dafermos, C.M., Nohel, J.A., A nonlinear hyperbolic Voltera equation in viscoelasticity, Amer. J. Math., 103 (Suppl.), (1981), 87–116.

[124] De Andrade, N.G., Feireisl, E., Well posedness for the problem of adiabatic shearing of fluids with discontinuous initial data, Nonlinear Anal., 20 (4), (1993), 389–394.

[125] Deeba, E., Khuri, S.A., Xie, S., An algorithm for solving a nonlinear integrodifferential equation, Appl. Math. Comput., 115, (2000), 123–131.

[126] Dehghan, M., Solution of a partial integro-differential equation arising from viscoelasticity, Int. J. Comput. Math., 83, (2006), 123–129.

[127] Dehghan, M., Shakeri, F., Solution of an integro-differential equation arising in oscillating magnetic fields using He's homotopy perturbation method, Progress in Electromagnetics Research, 78, (2008), 361–376.

[128] Dehghan, M., Shakeri, F., Solution of parabolic integro-differential equations arising in heat conduction in materials with memory via He's variational iteration technique, Int. J. Numer. Methods Biomed. Eng., 26, (2010), 705–715.

[129] Dendy, J.E., Galerkin's method for some highly nonlinear problems, SIAM J. Numer. Anal., 14, (1977), 327–347.

[130] Dickey, R.W., The initial value problem for a nonlinear semi-infinite string, Proc. Royal Soc. Edinburgh, Sec. A Math., 82, (1978), 19–26.

[131] Ding, S.J., Guo, B.L., Lin, J.Y., Zeng, M., Global existence of weak solutions for Landau–Lifshitz–Maxwell equations, Discrete Contin. Dyn. Syst., 17, (2007), 867–890.

[132] Douglas, J., Jones, B.F., Numerical methods for integro-differential equations of parabolic and hyperbolic types, Numer. Math., 4 (1), (1962), 96–102.

[133] Dubinskii, Yu. A., Nonlinear elliptic and parabolic equations, J. Math. Sci., 12 (5), (1979), 475–554.

[134] Dubinskii, Yu. A. Quasilinear elliptic and parabolic equations of arbitrary order, Uspekhi Mat. Nauk, 23, (1968), 45–90 (in Russian). English translation: Russ. Math. Surv., 23, (1968), 45–91.

[135] Duduchava, R., Integral Equations with Fixed Singularities. BG Teubner, 1979.

[136] Duvaut, G., Lions, J.-L., Inequalities in Mechanics and Physics, Springer-Verlag, Berlin, 1976.

[137] Dzhangveladze, T., An investigation of the First Boundary-Value Problem for Some Nonlinear Parabolic Integrodifferential Equations, Tbilisi State University, 1983 (in Russian).

[138] Dzhangveladze, T.A., First boundary value problem for a nonlinear equation of parabolic type, Dokl. Akad. Nauk SSSR, 269, (1983), 839–842 (in Russian). English translation: Soviet Phys. Dokl., 28, (1983), 323–324.

[139] Dzhangveladze, T.A., Investigation and Numerical Solution of Some Nonlinear Integro-differential Parabolic Problems, Candidate of Physical and Mathematical Sciences (Ph.D.) (in Russian), Diss. Bull. Candidate's Degree, Nikoloz Muskhelishvili Institute in Computational Mathematics of Georgian Academy of Sciences. Tbilisi, 1984 (in Georgian and Russian).

[140] Dzhangveladze, T.A., Solvability of the first boundary value problem for a nonlinear integro-differential equation of parabolic type, Bull. Acad. Sci. Georgian SSR (Soobshch. Akad. Nauk Gruz. SSR), 114, (1984), 261–264.

[141] Dzhangveladze, T.A., A nonlinear integro-differential equation of parabolic type, Differ. Uravn., 21 (1985), 41–46 (in Russian). English translation: Differ. Equ., 21, (1985), 32–36.

[142] Dzhangveladze, T.A., The difference scheme for one system of nonlinear partial differential equations, Rep. Enlarged Sess. Semin. I.Vekua Appl. Math., 2, (1986), 40–43 (in Russian).

[143] Dzhangveladze, T.A., Convergence of a difference scheme for a system of nonlinear partial differential equations, Soobshch. Akad. Nauk Gruz. SSR (Bull. Acad. Sci. Georgian SSR), 126 (2), (1987), 257–260 (in Russian).

[144] Dzhangveladze, T.A., Stability of the stationary solution of a system of nonlinear partial differential equations, Proc. All-Union Symp. Curr. Probl. Math. Phys., Tbilisi, 1, (1987), 214–221 (in Russian).

[145] Dzhangveladze, T.A., Mathematical Modelling, Investigation and Numerical Solution of Some Nonlinear Diffusion Problems, Doctoral Dissertation. Doctor of Physical and Mathematical Sciences (Dr. Habil) (in Russian). Diss. Bull. Doctor's Degree, Ivane Javakhishvili Tbilisi State University, Tbilisi, Georgia, 1998 (in Georgian and Russian).

[146] Dzhangveladze, T.A., Kiguradze, Z.V., On the stabilization of solutions of an initial-boundary value problem for a nonlinear integro-differential equation, Differ. Uravn., 43, (2007), 833–840 (in Russian). English translation: Differ. Equ., 43, (2007), 854–861.

[147] Dzhangveladze, T.A., Kiguradze, Z.V., Asymptotic behavior of the solution to nonlinear integro-differential diffusion equation, Differ. Uravn., 44, (2008), 517–529 (in Russian). English translation: Differ. Equ., 44, (2008), 538–550.

[148] Dzhangveladze, T.A., Lybimov, B.Y., Korshia, T.K., On the Numerical Solution of a Class of Nonisothermic Problems of the Diffusion of an Electromagnetic Field, Proc. I. Vekua Inst. Appl. Math. (Tbilisi. Gos. Univ. Inst. Prikl. Math. Trudy), 18, (1986), 5–47 (in Russian).

[149] Egorov, Yu.V., Galaktionov, V.A., Kondratiev, V.A., Pohožaev, S.I., Global solutions of higher-order semilinear parabolic equations in the supercritical range, Adv. Differential Equations, 9, (2004), 1009–1038.

[150] Egorov, Yu.V., Galaktionov, V.A., Kondratiev, V.A., Pohožaev, S.I., On the asymptotics of global solutions of higher-order semilinear parabolic equations in the supercritical range, C. R. Acad. Sci. Paris, Ser. 1, 335, (2002), 805–810.

[151] Eidelman, S.D., Parabolic Systems, North-Holland, Amsterdam, London, 1969.

[152] Eller, M., Lagnese, J.E., Nicaise, S., Stabilization of heterogeneous Maxwell's equations by linear or nonlinear boundary feedbacks, Electron. J. Differential Equations, 2002 (21), (2002), 1–26.

[153] Elliott, C.M., Larsson, S., A finite element model for the time-dependent Joule heating problem, Math. Comp., 64, (1995), 1433–1453.

[154] El-Sayed, S. M., Abdel-Aziz, M. R., A comparison of Adomian's decomposition method and wavelet-Galerkin method for solving integro-differential equations, Appl. Math. Comput., 136, (2003), 151–159.

[155] Engler, H., Global smooth solutions for a class of parabolic integrodifferential equations, Trans. Amer. Math. Soc., 348, (1996), 267–290.

[156] Engler, H., On some parabolic integro-differential equations: Existence and asymptotics of solutions, Proc. Int. Conf. Equadiff 82, Würzburg 1982, Lecture Notes Math., Vol. 1017, (1983), 161–167.

[157] Elschner, J., On spline approximation for a class of integral equations. I: Galerkin and collocation methods with piecewise polynomials, Math. Meth. Appl. Sci., 10, (1988), 543–559.

[158] Estep, D.J., Verduyn Lunel, S. M., Williams, R.D., Analysis of shear layers in a fluid with temperature-dependent viscosity, J. Comput. Physics, 173, (2001), 17–60.

[159] Estrada, R., Ram P. Kanwal, R.K., Singular integral equations, Springer, 2000.

[160] Evans, G.C., Sull'equazione integro-differenziale di tipo parabolico, Rend. R. Accad. dei Lincei Cl. Sci. Fis., Mat. Nat., ser. 5, 21, (1912), 25–31.

[161] Ewing, R., Lazarov, R., Lin, Y., Finite volume element approximations of nonlocal reactive flows in porous media, Numer. Methods Partial Differential Equations, 16 (3), (2000), 285–311.

[162] Fabrizio, M., Gentili, G., Reynolds D.W., On rigid linear heat conductors with memory, Internat. J. Engrg Sci., 36, (1998), 765–782.

[163] Fakhar-Izadi, F., Dehghan, M., The spectral methods for parabolic Volterra integro-differential equations, J. Comput. Appl. Math., 235, (2011), 4032–4046.

[164] Feller, W., On boundaries and lateral conditions for the Kolmogorov differential equations, Ann. of Math., 65 (3), (1957), 527–570.

[165] Fitzgibbon, W.E., Semilinear integrodifferential equations in a Banach space, Nonlinear Anal., 4, (1980), 745–760.

[166] Ford, N.J., Baker, C.T.H., Roberts, J.A., Nonlinear Volterra integro-differential equations - Stability and numerical stability of θ-methods, J. Integral Equations Appl., 10, (1997), 397–416.

[167] Forsythe, G.E., Wasow, W.R., Finite Difference Methods for Partial Differential Equations, John Wiley & Sons, Inc., New York, 1960.

[168] Fowler, A.C., Frigaard, I., Howison, S.D., Temperature surges in current-limiting circuit devices, SIAM J. Appl. Math., 52 (4), (1992), 998–1011.

[169] Friedman, A., Partial Differential Equations of Parabolic Type, Prentice-Hall, Englewood, NJ, 1964.

[170] Fujita, H., Mizutani, A., On the finite element method for parabolic equations. I. Approximation of holomorphic semigroups, J. Math. Soc. Japan, 28, (1976), 749–771.

[171] Fujita, H., Suzuki, T., On the finite element approximation for evolution equations of parabolic type, In: Computing Methods in Applied Sciences and Engineering, 1977. Third International Symposium, December 5-9, 1977, IRIA LABORIA, Institut de Recherche d'Informatique et d'Automatique, R. Glowinski, (ed.), Lecture Notes in Math., Vol. 704, Springer, Berlin, (1979), 207–221.

[172] Gagoshidze, M., Numerical resolution of one nonlinear parabolic system, Rep. Enlarged Sess. Semin. I. Vekua Appl. Math., 24, (2010), 40–44.

[173] Gagoshidze, M., On parabolic regularization for one nonlinear diffusion model, Rep. Enlarged Sess. Semin. I. Vekua Appl. Math., 26, (2012), 14–17.

[174] Gagoshidze, M., Jangveladze, T., On one nonlinear diffusion system, Rep. Enlarged Sess. Semin. I. Vekua Appl. Math., 25, (2011), 39–43.

[175] Gajewski, H., Gröger, K., Zacharias, K., Nichtlineare Operatorgleichungen und Operatordifferrentialgleichungen, Akademie-Verlag, Berlin, 1974.

[176] Galaktionov, V.A., Geometric Sturmian Theory of Nonlinear Parabolic Equations and Applications, Chapman & Hall/CRC Appl. Math. and Nonlinear Sci. Series, Boca Raton, London, New York, Washington, D.C., 2004.

[177] Garroni, M.G., Menaldi, J.-L., Green Functions for Second Order Parabolic Integro-Differential Problems, Pitman Research Notes in Mathematics Series Vol. 275, Longman, Essex, 1992.

[178] Ghasemi, M., Kajani, M. T., Babolian, E., Numerical solutions of the nonlinear integrodifferential equations: Wavelet-Galerkin method and homotopy perturbation method, Appl. Math. Comput., 188, (2007), 450–455.

[179] Giorgi, C., Gentili, G., Thermodynamic properties and stability for the heat flux equation with linear memory, Quart. Appl. Math., 51, (1993), 343–362.

[180] Giorgi, C., Marzocchi, A., Pata, V., Asymptotic behavior of a semilinear problem in heat conduction with memory, Nonlinear Differ. Equ. Appl., 5, (1998), 333–354.

[181] Glassey, R., Yin, H,-M., On Maxwell's equations with a temperature effect, Part II, Comm. Math. Phys., 194 (2), (1998), 343–358.

[182] Glowinski, R., Shiau, L., Sheppard, M. Numerical methods for a class of nonlinear integro-differential equations, Calcolo 50, (2013), 17–33.

[183] Godunov, S.K., Ryabenkii, V.S., Difference Schemes: An Introduction to the Underlying Theory, North-Holland,1987.

[184] Gohberg, I., Krupnik, I., Introduction to the Theory of One-Dimensional Singular Integral Operators, Ştiinţă, Chişinău, 1973 (in Russian), German translation: Birkhauser Verlag, Basel, 1979.

[185] Golovkin, K.K., New model equations of motion of a viscous fluid and their unique solvability, Trudy Mat. Inst. Steklov, 102, (1967), 29–50 (in Russian). English translation: Proc. Steklov Inst. Math., 102, (1970), 29 54.

[186] Gordeziani, D.G., Dzhangveladze, T.A., Korshia, T.K., A class of nonlinear diffusion equation, Proc. International Congress of Mathematicians, Warsaw, Poland, 16-24 August 1983, Sec. 11: Partial Differential Equations.

[187] Gordeziani, D.G., Dzhangveladze, T.A., Korshia, T.K., On a class of nonlinear parabolic equations arising in problems of the diffusion of an electromagnetic field, Proc. I. Vekua Inst.

Appl. Math. (Tbiliss. Gos. Univ. Inst. Prikl. Math. Trudy), 13, (1983), 7–35 (in Russian).

[188] Gordeziani, D.G., Dzhangveladze, T.A., Korshia, T.K., Existence and uniqueness of the solution of a class of nonlinear parabolic problems, Differ. Uravn., 19, (1983), 1197–1207 (in Russian). English translation: Differ. Equ., 19, (1984), 887–895.

[189] Gordis, J.H., Neta, B., An adaptive method for the numerical solution of Volterra integral equations, J. Vib. Acoust., 123, (2001), 545–547.

[190] Grigoriev, Y.N., Ibragimov, N.H., Kovalev, V.F., Meleshko, S.V., Symmetries of Integro-Differential Equations: With Applications in Mechanics and Plasma Physics, Springer, New York, 2010.

[191] Gripenberg, G., Global existence of solutions of Volterra integrodifferential equations of parabolic type, J. Differential Equations, 102, (1993), 382–390.

[192] Gripenberg, G., Londen, S.-O, Staffans, O., Volterra Integral and Functional Equations, Encyclopedia of Mathematics and Its Applications, Cambridge University Press, Cambridge, 1990.

[193] Grossmann, C., Roos, H.-G., Stynes, M., Numerical Treatment of Partial Differential Equations, Springer, Berlin, 2007.

[194] Grote, M.J., Local nonreflecting boundary condition for Maxwell's equations, Comput. Methods Appl. Mech. Engrg., 195, (2006), 3691–3708.

[195] Guo, Y., The Vlasov-Maxwell-Boltzmann system near Maxwellians, Invent. Math., 153, (2003), 593–630.

[196] Gurtin, M.E., Pipkin, A. C., A general theory of heat conduction with finite wave speeds, Arch. Ration. Mech. Anal., 31, (1968), 113–126.

[197] Hall, W.F., Kabakian, A.V., A sequence of absorbing boundary conditions for Maxwell's equations, J. Comput. Phys., 194, (2004), 140–155.

[198] Han, H., Zhu, L., Brunner, H., Ma, J.T., The numerical solution of parabolic Volterra integro-differential equations on unbounded spatial domains, Appl. Numer. Math., 55, (2005), 83–99.

[199] Haraux, A., Nonlinear Evolution Equations: Global Behavior of Solutions, Lecture Notes in Mathematics, Volume 841, Springer-Verlag, Berlin, 1981.

[200] Hashim, I., Adomian decomposition method for solving BVPs for fourth-order integrodifferential equations, J. Comput. Appl. Math., 193, (2006), 658–664.

[201] He, J.H., Homotopy perturbation technique, Comput. Methods Appl. Mech. Engrg., 178, (1999), 257–262.

[202] He, J.H., Homotopy perturbation method: a new nonlinear analytical technique, Appl. Math. Comput., 135, (2003), 73–79.

[203] Heard, M.L., An abstract parabolic Volterra integro-differential equation, SIAM J. Math. Anal., 13, (1982), 81–105.

[204] Heard, M., Rankin, S.M., A semilinear parabolic Volterra integrodifferential equation, J. Differential Equations, 71, (1988), 201–233.

[205] Heise, B., Analysis of a fully discrete finite element method for a nonlinear magnetic field problem, SIAM J. Numer. Anal., 31, (1994), 745–759.

[206] Herrmann, R.P., Nachlinger, R., On uniqueness and wave propagation in nonlinear heat conductors with memory, J. Math. Anal. Appl., 50, (1975), 530–547.

[207] Hernandez, C., Ovando-Martinez, R.B.B., Arjona, M.A., Application of tensor analysis to the finite element method, Appl. Math.Comput., 219 (9), (2013), 4625–4636.

[208] Hong, M.-C., Tonegawa, Y., Yassin, A., Partial regularity of weak solutions to Maxwell's equations in a quasi-static electromagnetic field, Methods Appl. Anal., 15, (2008), 205–222.

[209] Howison, S.D., Rodrigues, J.F., Shillor, M., Stationary solutions to the thermistor problem, J. Math. Anal. Appl., 174 (2), (1993), 573–588.

[210] Hundsdorfer, W., Verwer, J.G., Numerical Solution of Time-Dependent Advection-Diffusion-Reaction Equations, Springer Series in Comp. Math., Volume 33, 2003.

[211] Hutton, D.V., Fundamentals of Finite Element Analysis, McGraw-Hill, New York, 2004.

[212] Hyman, J.M., Shashkov, M., Mimetic discretizations for Maxwells equations, J. Comput. Phys., 151, (1999), 881–909.

[213] Iliev, O.P., Margenov, S.D., Minev, P.D., Vassilevski, P.S., Zikatanov, L.T., Numerical Solution of Partial Differential Equations: Theory, Algorithms, and Their Applications, Springer, 2013.

[214] Iskakov, A.B., Descombes, S., Dormy, E., An integro-differential formulation for magnetic induction in bounded domains: boundary element-finite volume method, J. Comput. Phys., 197, (2004), 540–554.

[215] Ivanov, V., The Theory of Approximative Methods and their Application to the Numerical Solution of Singular Integral Equations, Noordhoff, Leyden, The Netherlands, 1976.

[216] Jakobsen, E.R., Karlsen, K.H., Continuous dependence estimates for viscosity solutions of integro-PDEs, J. Differential Equations, 212, (2005), 278–318.

[217] Jangveladze, T., Apriori estimations for one nonlinear integro-differential parabolic problem, Rep. Enlarged Sess. Semin. I. Vekua Appl. Math., 8, (1993), 35–37.

[218] Jangveladze, T., Asymptotic behavior ($t \to \infty$) of solutions for one nonlinear integro-differential parabolic equation, Rep. Enlarged Sess. Semin. I. Vekua Appl. Math., 12, (1997), 9–11.

[219] Jangveladze, T., On one class of nonlinear integro-differential parabolic equations, Semin. I. Vekua Inst. Appl. Math. Rep., 23, (1997), 51–87.

[220] Jangveladze, T., Convergence of a difference scheme for a nonlinear integro-differential equation, Proc. I. Vekua Inst. Appl. Math., 48, (1998), 38–43.

[221] Jangveladze, T., Additive models for one nonlinear diffusion system, Rep. Enlarged Sess. Semin I. Vekua Appl. Math., 24 (2010), 66–69.

[222] Jangveladze, T. Asymptotic properties of solution for one nonlinear integro-differential model associated with the penetration of a magnetic field into a substance, First Int. Conf. Georgian Math. Union. Book of Abstracts, (2010), 80–81.

[223] Jangveladze, T., Investigation and numerical solution of system of nonlinear integro-differential equations associated with the penetration of a magnetic field in a substance, Recent Researches in Applied Mathematics, Proc. 15th WSEAS Int. Conf. Applied Math. (MATH '10), Vouliagmeni, Athens, Greece, December 29-31, 2010, 79–84.

[224] Jangveladze, T., On the parabolic regularization of one nonlinear diffusion system, II International Conference of Georgian Mathematical Union, (2011), 129.

[225] Jangveladze, T., Some properties and numerical solution of one-dimensional nonlinear electromagnetic diffusion system, Advances in Applied and Pure Mathematics, (2014), 96–100.

[226] Jangveladze, T., Some properties of solutions and approximate algorithms for one system of nonlinear partial differential equations, International Workshop on the Qualitative Theory of Differential Equations, QUALITDE - 2014, (2014), 54–57.

[227] Jangveladze, T., Aptsiauri, M., The asymptotic behavior of solutions and finite difference scheme for one nonlinear integro-differential equation, Bull. Georgian Acad. Sci., 174, (2006), 25–28.

[228] Jangveladze, T.A., Kiguradze, Z., The asymptotic behavior of the solutions of one nonlinear integro-differential parabolic equation, Rep. Enlarged Sess. Semin. I. Vekua Appl. Math., 10, (1995), 36–38.

[229] Jangveladze, T., Kiguradze, Z., On the asymptotic behavior of solution for one system of nonlinear integro-differential equations, Rep. Enlarged Sess. Semin. I. Vekua Appl. Math., 14, (1999), 35–38.

[230] Jangveladze, T.A., Kiguradze, Z.V., On the difference schemes for one nonlinear diffusion model, Semin. I. Vekua Inst. Appl. Math. Rep., 25, (1999), 89–95.

[231] Jangveladze, T., Kiguradze, Z., Estimates of a stabilization rate as $t \to \infty$ of solutions of a nonlinear integro-differential equation, Georgian Math. J., 9, (2002), 57–70.

[232] Jangveladze, T., Kiguradze, Z., Estimates of a stabilization rate as $t \to \infty$ of solutions of a nonlinear integro-differential diffusion system, Appl. Math. Inform. Mech., 8, (2003), 1–19.

[233] Jangveladze, T., Kiguradze, Z., Estimates of a stabilization rate as $t \to \infty$ of solutions of a system of nonlinear integro-

differential equations, Rep. Enlarged Sess. Semin. I. Vekua Appl. Math., 18, (2003), 16–19.

[234] Jangveladze, T., Kiguradze, Z., On one nonlinear integro-differential diffusion system, Bull. Georgian Acad. Sci., 170, (2004), 42–45.

[235] Jangveladze, T., Kiguradze, Z., The asymptotic behavior as $t \to \infty$of the solution of one nonlinear integro-differential equation, Bull. Georgian Acad. Sci., 170, (2004), 228–230.

[236] Jangveladze, T., Kiguradze, Z., Long time behavior of solutions to nonlinear integro-differential equation, Proc. I. Vekua Inst. Appl. Math. (Tr. Inst. Prikl. Mat. Im. I.N.Vekua), 54-55, (2004-2005), 65–73.

[237] Jangveladze, T., Kiguradze, Z., On the asymptotic behaviour as $t \to \infty$ of solutions of one nonlinear integro-differential parabolic equation arising in penetration of a magnetic field into a substance, Rep. Enlarged Sess. Semin. I. Vekua Appl. Math., 20, (2005), 8–11.

[238] Jangveladze, T.A., Kiguradze, Z.V., Asymptotics of a solution of a nonlinear system of diffusion of a magnetic field into a substance, Sibirsk. Mat. Zh., 47, (2006), 1058–1070 (in Russian). English translation: Siberian Math. J., 47, (2006), 867–878.

[239] Jangveladze, T., Kiguradze, Z., On nonlinear integro-differential equations based on Maxwell's system, Proc. I. Vekua Inst. Appl. Math., 56-57, (2006-2007), 55–65.

[240] Jangveladze, T., Kiguradze, Z., Large time asymptotics of solutions to a nonlinear integro-differential equation, Mem. Diff. Eq. Math. Phys., 42, (2007), 35–48.

[241] Jangveladze, T., Kiguradze, Z., Large time behavior of solutions and difference schemes to nonlinear integro-differential

system associated with the penetration of a magnetic field into a substance, Appl. Math. Inform. Mech., 13, (2008), 40–54.

[242] Jangveladze, T., Kiguradze, Z., Large time behavior of solutions to one nonlinear integro-differential equation, Georgian Math. J., 15, (2008), 531–539.

[243] Jangveladze, T., Kiguradze, Z., Asymptotics of solution and finite difference scheme to a nonlinear integro-differential equation associated with the penetration of a magnetic field into a substance, WSEAS Trans. Math., 8, (2009), 467–477.

[244] Jangveladze, T., Kiguradze, Z., Finite difference scheme to a nonlinear integro-differential equation associated with the penetration of a magnetic field into a substance, Proc. 2nd WSEAS Int. Conf. Finite Differences, Finite Elements, Finite Volumes, Boundary Elements (F-and-B '09), (2009), 186–192.

[245] Jangveladze, T., Kiguradze, Z., Asymptotics for large time of solutions to nonlinear system associated with the penetration of a magnetic field into a substance, Appl. Math., 55, (2010), 471–493.

[246] Jangveladze, T., Kiguradze, Z., Large time behavior of the solution to an initial-boundary value problem with mixed boundary conditions for a nonlinear integro-differential equation, Cent. Eur. J. Math., 9, (2011), 866–873.

[247] Jangveladze, T., Kiguradze, Z., On some nonlinear partial differential and integro-differential diffusion models, WSEAS, Proc. 2nd Intern. Conf. Appl., Numer. Comput. Math., (2013), 124–129.

[248] Jangveladze, T., Kiguradze, Z., On investigation and approximate solution of one nonlinear partial integro-differential equation with source term, Recent Advances in Applied Mathematics, Modelling and Simulation, (2014), 50–55.

[249] Jangveladze, T., Kiguradze, Z., On one nonlinear integro-differential equation with source term, Rep. Enlarged Sess. Semin. I.Vekua Appl. Math., 27, (2013), 23–26.

[250] Jangveladze, T., Kiguradze, Z., Semi-discrete scheme for one nonlinear integro-differential system describing diffusion process of electromagnetic field, Advances in Applied and Pure Mathematics, (2014), 118–122.

[251] Jangveladze, T., Kratsashvili, M., Asymptotic behavior of solution and semi-discrete difference scheme for one nonlinear integro-differential equation with source term, Rep. Enlarged Sess. Semin. I.Vekua Appl. Math., 28, (2014), 50–53.

[252] Jangveladze, T., Kiguradze, Z., Gagoshidze, M., Large time behavior of solution and semi-discrete scheme for one nonlinear integro-differential equation with source terms, J. Appl. Math. Inform. Mech., 19, (2014), 10–17.

[253] Jangveladze, T., Kiguradze, Z., Kratsashvili, A., Asymptotic behavior of the solution and semi-discrete finite difference scheme for one nonlinear integro-differential model with source terms, J. Appl. Math. Inform. Mech., 18, (2013), 19–30.

[254] Jangveladze, T., Kiguradze, Z., Neta, B., Large time behavior of solutions to a nonlinear integro-differential system, J. Math. Anal. Appl., 351, (2009), 382–391.

[255] Jangveladze, T., Kiguradze, Z., Neta, B., Finite difference approximation of a nonlinear integro-differential system, Appl. Math. Comput., 215, (2009), 615–628.

[256] Jangveladze, T., Kiguradze, Z., Neta, B., Large time behavior of solutions and finite difference scheme to a nonlinear integro-differential equation, Comput. Math. Appl., 57, (2009), 799–811.

[257] Jangveladze, T., Kiguradze, Z., Neta, B., Large time asymptotic and numerical solution of a nonlinear diffusion model with memory, Comput. Math. Appl., 59, (2010), 254–273.

[258] Jangveladze, T., Kiguradze, Z., Neta, B., Galerkin finite element method for one nonlinear integro-differential model, Appl. Math. Comput., 217, (2011), 6883–6892.

[259] Jangveladze, T., Kiguradze, Z., Neta, B., Finite element method for a system of nonlinear integro-differential equations with mixed boundary conditions, Rep. Enlarged Sess. Semin. I. Vekua Appl. Math., 26, (2012), 23–28.

[260] Jangveladze, T., Kiguradze, Z., Gagoshidze, M., Nikolishvili, M., Stability and convergence of the variable directions difference scheme for one nonlinear two-dimensional model, Internat. J. Biomath., 8, (2015), 1550057, 21 pages.

[261] Jangveladze, T., Kiguradze, Z., Neta, B., Reich, S., Finite element approximations of a nonlinear diffusion model with memory, Numer. Algor., 64, (2013), 127–155.

[262] Jia, S., Li, D., Liu, T., Zhang, S.H., Richardson extrapolation and defect correction of mixed finite element methods for integro-differential equations in porous media, Appl. Math., 53, (2008), 13–39.

[263] Jiang, Z., $L^\infty(L^2)$ and $L^\infty(L^\infty)$ Error estimates for mixed methods for integro-differential equations of parabolic type, ESAIM: Mathematical Modelling and Numerical Analysis (M2AN), 33, (1999), 531–546.

[264] Kachurovskii, R.I., Non-linear monotone operators in Banach spaces, Russian Math. Surveys, 23, (1968), 117–165.

[265] Kalinin, Y.N., Reitmann, V., Yumaguzin, N.Y., Asymptotic behavior of Maxwell's equation in one-space dimension with thermal effect, Discrete Dyn. Syst. Suppl., 2, (2011), 754–762.

[266] Kang, K., Kim, S., On the Hölder continuity of solutions of a certain system related to Maxwell's equations, SIAM J. Math. Anal., 34 (1), (2002), 87–100.

[267] Kang, K., Kim, S., Minut, A., On the regularity of solutions to a parabolic system related to Maxwell's equations, J. Math. Anal. Appl., 299 (1), (2004), 89–99.

[268] Kačur, J., Application of Rothe's method to evolution integrodifferential equations, J. Reine Angew. Math., 388, (1988), 73–105.

[269] Kauthen, J.-P., The method of lines for parabolic partial integro-differential equations, J. Integral Equations Appl., 4, (1992), 69–81.

[270] Kazhikhov, A. V., On stabilization of solutions of the initial-boundary value problem for the equations of a barotropic viscous fluid equations, Differ. Uravn., 15, (1979), 662–667 (in Russian). English translation: Differ. Equ., 15, (1979), 463–467.

[271] Kerimov, M. K., Translation of : Theory of Functionals and of Integral and Integro Differential Equations by V. Volterra, Nauka, Moscow, 1982.

[272] Khan, M., A new algorithm for higher order integro-differential equations, Afrika Mat., 26, (2015), 247–255.

[273] Kiguradze, Z., The difference scheme for one system of nonlinear partial differential equations, Rep. Enlarged Sess. Semin I. Vekua Appl. Math., 14 (1999), 67–70.

[274] Kiguradze, Z., Finite difference scheme for a nonlinear integro-differential system, Proc. I. Vekua Inst. Appl. Math., 50-51, (2000-2001), 65–72.

[275] Kiguradze, Z., On the stationary solution for one diffusion model, Rep. Enlarged Sess. Semin. I. Vekua Appl. Math., 16, (2001), 17–20.

[276] Kiguradze, Z., Investigation and Numerical Solution for One Nonlinear Diffusion Model, (Georgian, Russian) Candidate of Physical and Mathematical Sciences (Ph.D.), 2002 (in Georgian), Diss. Bull. Candidate's Degree, Ivane Javakhishvili Tbilisi State University, Tbilisi, Georgia, 2003 (in Georgian and Russian).

[277] Kiguradze, Z., Asymptotic behavior and numerical solution of the system of nonlinear integro-differential equations, Rep. Enlarged Sess. Semin. I. Vekua Appl. Math., 19, (2004), 58–61.

[278] Kiguradze, Z., The asymptotic behavior of the solutions of one nonlinear integro-differential model, Semin. I. Vekua Inst. Appl. Math. Rep., 30, (2004), 21–32.

[279] Kiguradze, Z., Numerical resolution of initial-boundary value problem for one nonlinear integro-differential equation, First Int. Conf. Georgian Math. Union. Book of Abstracts, (2010), 84–85.

[280] Kiguradze, Z., On asymptotic behavior and numerical resolution of one nonlinear Maxwell's model, Recent Researches in Applied Mathematics, Proc. 15th WSEAS Int. Conf. Applied Math. (MATH '10), Vouliagmeni, Athens, Greece, December 29-31, 2010, (2010), 55–60.

[281] Kiguradze, Z., Galerkin method for one nonlinear integro-differential equation with mixed boundary conditions, Appl. Math. Inform. Mech., 17, (2012), 31–38.

[282] Kiguradze, Z., Investigation and numerical solution of some systems of partial integro-differential equations, International Workshop on the Qualitative Theory of Differential Equations, QUALITDE - 2014, (2014), 81–85.

[283] Kiguradze, Z., Aptsiauri, M., Large time behavior of the solutions and finite difference scheme for one system of nonlinear integro-differential equations, Bull. Georgian Acad. Sci., 174, (2006), 211–214.

[284] Kiguradze, Z., Kratsashvili, A., Semi-discrete scheme for one system of nonlinear integro-differential equations with source terms, Rep. Enl. Sess. Sem. of I.Vekua Inst. Appl. Math., 28, (2014), 62–65.

[285] Kim, C.H., Choi, U.J., Spectral collocation methods for a partial integro-differential equation with a weakly singular kernel, J. Austral. Math. Soc. Ser. B, 39, (1998), 408–430.

[286] Kirchhoff, G., Vorlesungen über Mechanik, Teubner, Leipzig, 1876.

[287] Kokilashvili, V., Samko, S., Singular integrals in weighted Lebesgue spaces with variable exponent, Georgian Math. J., 10, (2003), 145–156.

[288] Kolmogorov, A.N., On Analytic Methods in Probability Theory, Selected Works of A. N. Kolmogorov, Volume II: Probability Theory and Mathematical Statistics (Ed. A. N. Shiryaev), Kluwer, Dordrecht, 1992, 62–108.

[289] Krasnosel'skii, M.A., Zabreiko, P.P., Geometrical Methods of Nonlinear Analysis. Vol. 263. Springer, 1984.

[290] Krejčí, P., Periodic solutions to Maxwell equations in nonlinear media, Czechoslovak Math. J., 36, (1986), 238–258.

[291] Kupradze, V.D., To the theory of integral equations with the integral in the sense of the Cauchy principal value, Soobshch. Akad. Nauk Gruz. SSR (Bull. Acad. Sci. Georgian SSR), 2, (1941), 587–596 (in Russian).

[292] Kythe, P.K., Puri, P., Computational Methods for Linear Integral Equations, University of New Orleans, New Orleans, 2002.

[293] Ladopoulos, E.G., Zisis V.A., Existence and uniqueness for non-linear singular integral equations used in fluid mechanics, Appl. Math., 42 (5), (1997), 345–367.

[294] Ladopoulos, E., Singular Integral Equations: Linear and Nonlinear Theory and its Applications in Science and Engineering, Springer, Berlin, Heidelberg, New York, 2000.

[295] Ladyzhenskaya, O.A., New equations for the description of the motions of viscous incompressible fluids, and global solvability for their boundary value problems, Trudy Mat. Inst. Steklov, 102, (1967), 85–104 (in Russian). English translation: Proc. Steklov Inst. Math., 102, (1967), 95–118.

[296] Ladyzhenskaya, O.A., The Mathematical Theory of Viscous Incompressible Flow, Gordon & Breach, Second ed., New York, 1969.

[297] Ladyzenskaya, O.A., Solonnikov, V.A., Solution of some nonstationary problems of magnetohydrodynamics for a viscous incompressible fluid, Mathematical problems of hydrodynamics and magnetohydrodynamics for a viscous incompressible fluid, Collected papers, Trudy Mat. Inst. Steklov., 59, (1960), 115–173 (in Russian). English translation: Proc. Steklov Math. Inst., 59, (1960), 115–173.

[298] Ladyzenskaya, O.A., Solonnikov, V.A., Ural'ceva, N.N., Linear and Quasilinear Equations of Parabolic Type, Transl. Math. Monographs, Volume 23, Amer. Math. Soc. , Providence, RI, 1968.

[299] Lakshmikantham, V., Rao, M.R.M., Theory of Integro-Differential Equations, CRC Press, 1995.

[300] Landau, L., Lifschitz, E., Electrodynamics of Continuous Media, Course of Theoretical Physics, 8 (Translated from the Russian), Pergamon Press, Oxford-London-New York-Paris, Addison-Wesley Publishing Co., Inc., Reading, Mass., 1960; Russian original: Gosudarstv. Izdat. Tehn-Teor. Lit., Moscow, 1957.

[301] Lapidus. L., Pinder. G.F., Numerical Solution of Partial Differential Equations in Science and Engineering, John Wiley & Sons, 1999.

[302] Laptev, G.I., Conditions for the solvability of a problem on the diffusion of a magnetic field in matter taking into account the heating of the body, Mech. Appl. Math. (Priok. Knizhn. Izdat., Tula), 1988, 126–131 (in Russian).

[303] Laptev, G.I., Mathematical features of the penetration of magnetic field into mater in quasistationary approximation, Izv. Vyssh. Uchebn. Zaved. Mat., 33 (7), (1989), 70–73 (in Russian). English translation: Soviet Math., 33 (7), (1989), 93–97.

[304] Laptev, G.I., Mathematical features of the problem of the penetration of a magnetic field into matter, Zh. Vychisl. Mat. Mat. Fiz., 28, (1988), 1332–1345. English translation: U.S.S.R. Comput. Math. Math. Phys., 28, (1990), 35–45.

[305] Laptev, G.I., Quasilinear parabolic equations which contains in coefficients Volterra's operator, Math. Sbornik, 136, (1988), 530–545 (in Russian), English translation: Sbornik Math., 64, (1989), 527-542.

[306] Laptev, G.I., Quasilinear Evolution Partial Differential Equations with Operator Coefficients, Doctoral Dissertation, M. V. Keldish Inst. Appl. Math. Russian Acad. Sci., Moscow, 1990 (in Russian).

[307] Laptev, G. I., Degenerate quasilinear evolution equations containing a Volterra operator in the coefficients, Arabian J. Sci. Eng., Sect. B Eng., 17, (1992), 591–598.

[308] Larsson, S., Thomée, V., Wahlbin, L.B., Numerical solution of parabolic integro-differential equations by the discontinuous Galerkin method, Math. Comp., 67, (1998), 45–71.

[309] Lasiecka, I., Ong, J., Global solvability and uniform decays of solutions to quasilinear equations with nonlinear boundary dissipation, Comm. Partial Differential Equations, 24, (1999), 2069–2107.

[310] Lavrentiev, M.M., Jr., Spigler, R., Existence and uniqueness of solutions to the Kuramoto-Sakaguchi nonlinear parabolic integrodifferential equation, Differential Integral Equations, 13, (2000), 649–667.

[311] Lavrentiev, M.M., Jr., Spigler, R., Akhmetov, D.R., Nonlinear integroparabolic equations on unbounded domains: Existence of classical solutions with special properties, Siberian Math. J., 42, (2001), 495–516.

[312] Lavrentiev, M. M., Jr., Spigler, R., Akhmetov, D. R., Regularizing a nonlinear integroparabolic Fokker-Planck equation with space-periodic solutions: Existence of strong solutions, Siberian Math. J., 42, (2001), 693–714

[313] Le Roux, M.N., Thomée, V., Numerical solution of semilinear integrodifferential equations of parabolic type with nonsmooth data, SIAM J. Numer. Anal., 26, (1989), 1291–1309.

[314] Li, J., Error analysis of finite element methods for 3-D Maxwell's equations in dispersive media, J. Comput. Appl. Math., 188, (2006), 107–120.

[315] Li. J., Error analysis of fully discrete mixed finite element schemes for 3-D Maxwell's equations in dispersive media, Comput. Methods Appl. Mech. Engrg., 196, (2007), 3081–3094.

[316] Li, J., Posteriori error estimation for an interior penalty discontinuous Galerkin method for Maxwell's equations in cold plasma, Adv. Appl. Math. Mech., 1, (2009), 107–124.

[317] Li, J., Chen, Y., Analysis of a time-domain finite element method for 3-D Maxwell's equations in dispersive media, Comput. Methods Appl. Mech. Engrg., 195, (2006), 4220–4229.

[318] Liao, H., Zhao, Y., Linearly localized difference schemes for the nonlinear Maxwell model of a magnetic field into a substance, Appl. Math. Comput., 233, (2014), 608–622.

[319] Lin, Y., Galerkin methods for nonlinear parabolic integrodifferential equations with nonlinear boundary conditions, SIAM J. Numer. Anal., 27, (1990), 608–621.

[320] Lin, Y., Semi-discrete finite element approximations for linear parabolic integro-differential equations with integrable kernels, J. Integral Equations Appl., 10, (1998), 51–83.

[321] Lin, Y., Thomée, V., Wahlbin, L.B., Ritz-Volterra projections to finite-element spaces and applications to integrodifferential and related equations, SIAM J. Numer. Anal., 28, (1991), 1047–1070.

[322] Lin, Y., Yin, H.-M., Nonlinear parabolic equations with nonlinear functionals, J. Math. Anal. Appl., 168, (1992), 28–41.

[323] Lin, Y.P., On maximum norm estimates for Ritz-Volterra projection with applications to some time dependent problems, J. Comput. Math, 15 (1997), 159–178.

[324] Lin, Q., Zhang, S., Yan, N., Methods for improving approximate accuracy for hyperbolic integrodifferential equations, Systems Sci. Math. Sci., 10, (1997), 282–288.

[325] Lin, Q., Zhang, S., Yan, N., High accuracy analysis for integrodifferential equations, Acta Math. Appl. Sin., 14 (2), (1998), 202–211.

[326] Linz, P., Analytical and Numerical Methods for Volterra Equations, SIAM, Philadelphia, PA, 1985.

[327] Lions, J.-L., Quelques Méthodes de Résolution des Problèmes aux Limites Non-Linéaires, Dunod, Gauthier-Villars, Paris, 1969 (in French).

[328] Lions, J.-L., On Some Questions in Boundary Value Problems of Mathematical Physics, Contemporary Developments in Continuum Mechanics and Partial Differential Equations, Proc. Internat. Sympos., Inst. Math. Univ. Fed. Rio de Janeiro, 1977, 284–346.

[329] Lions, P.L., On some recent methods for nonlinear partial differential equations, Proc. International Congr. Math., Zurich I, (1994), 140–155.

[330] Litvinchuk, G., Solvability Theory of Boundary Value Problem and Singular Integral Equations with Shift, Mathematics and its Applications, Kluwer Academic Publishers, 2000.

[331] Liu, S,. Rincon, M.A., Effect of moving boundaries on the vibrating elastic string, Appl. Numer. Math., 47, (2003), 159–172.

[332] Londen, S.O., Some existence results for a nonlinear hyperbolic integrodifferential equation with singular kernel, J. Integral Equations Appl., 3, (1991), 3–30.

[333] Long, N.T., Dinh, A.P.N., Nonlinear parabolic problem associated with the penetration of a magnetic field into a substance, Math. Methods Appl. Sci., 16, (1993), 281–295.

[334] Long, N.T., Dinh, A.P.N., Periodic solutions of a nonlinear parabolic equation associated with the penetration of a magnetic field into a substance, Comput. Math. Appl., 30, (1995), 63–78.

[335] López-Marcos, J.C., A difference scheme for a nonlinear partial integrodifferential equation, SIAM J. Numer. Anal., 27, (1990), 20–31.

[336] Lu, T., Zhang, P., Cai, W., Discontinuous Galerkin methods for dispersive and lossy Maxwell's equations and PML boundary conditions, J. Comput. Phys., 200, (2004), 549–580.

[337] Lubich, C., Ostermann, A., Runge–Kutta approximation of quasi-linear parabolic equations, Math. Comp., 64, (1995), 601–627.

[338] Lunardi, A., Sinestrari, E., Existence in the large and stability for nonlinear Volterra equations, in "Integro-differential evolution equations and applications" (Trento, 1984), J. Integral Equations, 10, (1985), 213–239.

[339] Lunardi, A., Sinestrari, E., Fully nonlinear integrodifferential equations in general Banach spaces, Math. Z., 190, (1985), 225–248.

[340] Lunardi, A., Sinestrari, E., C-regularity for non-autonomous linear integrodifferential equations of parabolic type, J. Differential Equations, 63 (1), (1986), 88–116.

[341] Ma, C., Finite-element method for time-dependent Maxwell's equations based on an explicit magnetic-field scheme, J. Comput. Appl. Math., 194, (2006), 409–424.

[342] Ma, J., Brunner, H., A posteriori error estimates of discontinuous Galerkin methods for non-standard Volterra integro-differential equations, IMA J. Numer. Anal., 26, (2006), 78–95.

[343] Ma, J., Jiang, Y., Xiang, K., On a moving mesh method for solving partial integro-differential equations, J. Comput. Math., 27, (2009), 713–728.

[344] MacCamy, R.C., Stability theorems for a class of functional differential equations, SIAM J. Appl. Math., 30, (1976), 557–576.

[345] MacCamy, R.C., An integro-differential equation with application in heat flow, Quart. Appl. Math., 35, (1977), 1–19.

[346] MacCamy, R.C., A model for one-dimensional, nonlinear viscoelasticity, Quart. Appl. Math., 35, (1977), 21–23.

[347] Magnaradze, L., On one integral equation of the aeroplane wing theory, Soobshch. Acad. Nauk Gruzin. SSR, 3, (1942), 503–508.

[348] Maleknejad, K., Basirat, B., Hashemizadeh, E., A Bernstein operational matrix approach for solving a system of high order linear Volterra–Fredholm integro-differential equations, Math. Comput. Modelling, 55, (2012), 1363–1372.

[349] Manfrin, R., Global solvability to the Kirchhoff equation for a new class of initial data, Port. Math., 59, (2002), 91–109.

[350] Mastorakis, N.E., Numerical schemes for non-linear problems in fluid mechanics, Proc. 4th IASME/WSEAS International Conference on Continuum Mechanics, Cambridge, UK, (2009), 56–61.

[351] Matejíčka, L., On approximate solutions of degenerate integrodifferential parabolic problems, Math. Slovaca, 45, (1995), 91–103.

[352] Matsumura, A., Global existence and asymptotics of the solutions of the second-order quasilinear hyperbolic equations with first order dissipation, Publ. RIMS, Kyoto Univ., 13, (1977), 349–379.

[353] McLean, W., Sloan, I.H., Thomée, V., Time discretization via Laplace transformation of an integro-differential equation of parabolic type, Numer. Math., 102, (2006), 497–522.

[354] McLean, W., Thomée, V., Numerical solution of an evolution equation with a positive-type memory term, J. Austr. Math. Soc., Ser. B, Appl. Math., 35, (1993), 23–70.

[355] Medeiros, L. A., On a new class of nonlinear wave equations, J. Math. Anal. Appl., 69, (1979), 252–262.

[356] Medeiros, L.A., Ferrel, J.L., Menezes, S.B., Vibrations of elastic strings: Mathematical aspects, Part I., J. Comput. Anal. Appl., 4, (2002), 91–127

[357] Medeiros L.A., Ferrel J.L., Menezes S.B., Vibrations of elastic strings: Mathematical aspects, Part II., J. Comput. Anal. Appl., 4, (2002), 211–263.

[358] Melton, T.G., Vatsala, A.S., Higher order convergence for nonlinear parabolic integro-differential equations by generalized quasilinearization, Applications of Math. in Engineering and Economics: Proc. 34th Conf. on Applications of Mathematics in Engineering and Economics (AMEE '08), AIP Conf. Proc. Vol. 1067, (2008), 381–389.

[359] Metaxas, A.C., Foundations of Electroheat, A Unified Approach, John Wiley & Sons, New York, 1996.

[360] Metaxas, A.C., Meredith, R.J., Industrial Microwave Heating, IEE Power Engineering Series, Vol. 4, Peter Peregrinus Ltd., London, U.K., 1983.

[361] Mikhlin, S., Prossdorf, S., Singular Integral Operators, Springer-Verlag, Berlin, 1986.

[362] Milani, A., Local in time existence for the complete Maxwell equations with monotone characteristic in a bounded domain, Ann. Mat. Pura Appl., 131, (1982), 233–254.

[363] Miller, R.K., An integrodifferential equation for rigid heat conductors with memory, J. Math. Anal. Appl., 66, (1978), 313–332.

[364] Minty, G.J., Monotone (nonlinear) operators in Hilbert space, Duke Math. J., 29, (1962), 341–346.

[365] Miranda, F., Rodrigues, J.-F., Santos, L., A class of stationary nonlinear Maxwell systems, Math. Models Methods Appl. Sci., 19, (2009), 1883–1905.

[366] Mirzaee, F., Bimesl, S., Application of Euler matrix method for solving linear and a class of nonlinear Fredholm integro-differential equations, Mediterr. J. Math., 11, (2014), 999–1018.

[367] Mitrea, M., Boundary value problems for Dirac operators and Maxwell's equations in non-smooth domains, Math. Methods Appl. Sci., 25, (2002), 1355–1369.

[368] Mohsen, A., El-Gamel, M., A Sinc-Collocation method for the linear Fredholm integro-differential equations, Z. Angew. Math. Phys., 58 (3), (2007), 380–390.

[369] Monk, P., Finite Element Methods for Maxwell's Equations, Oxford University Press, New York, 2003.

[370] Morgan, J., Yin, H-M., On Maxwell's system with a thermal effect, Discrete Contin. Dyn. Syst. Ser. B., 1 (4), (2001), 485–494.

[371] Morton, K.W., Mayers, D.F., Numerical Solution of Partial Differential Equations, Cambridge University Press, New York, 1994.

[372] Muskelishvili, N., Singular Integral Equations. Boundary Problems of Function Theory and their Application to Mathematical Physics, Noordhoff, Leyden, The Netherlands, 1977.

[373] Nash, W.A., Modeer, J.R., Certain approximate analyses of the nonlinear behavior of plates and shallow shells, Proceedings of IUTAM Symposium on Theory of Thin Elastic Shells, North Holland Publishing Company, Amsterdam (1960), 331–354.

[374] Natanson, I.P., The Theory of Functions of a Real Variable, Nauka, Moscow, 1960 (in Russian). English translation: Theory of Functions of a Real Variable, Ney York, 1964.

[375] Neta, B., Finite element approximation of a nonlinear parabolic problem, Comput. Math. Appl., 4, (1978), 247–255.

[376] Neta, B., Numerical solution of a nonlinear integro-differential equation, J. Math. Anal. Appl., 89, (1982), 589–611.

[377] Neta, B., Adaptive method for the numerical solution of Fredholm integral equations of the second kind. Part I: Singular kernel, in Numerical Solution of Singular Integral Equations (A.Gerasoulis and R.Vichnevetsky, Eds.), (1984), 68–72.

[378] Neta, B., Igwe, J.O., Finite differences versus finite elements for solving nonlinear integro-differential equations, J. Math. Anal. Appl., 68, (1985), 607–617.

[379] Neta, B., Nelson, P., Adaptive method for the numerical solution of Fredholm integral equations of the second kind. Part II: Regular kernels, Appl. Math. Comput., 21, (1987), 171–184.

[380] Newman, W., Global solution of a nonlinear string equations, J. Math. Anal. Appl., 192, (1995), 689–704.

[381] Nirenberg, L., Partial differential equations in the first half of the century, in Development of Mathematics 1900-1950 (J. P. Pier, Ed.), Birkhäuser, Basel, 1994.

[382] Nishida, T., Global smooth solutions for the second order quasilinear wave equation with first order dissipation, unpublished note, (1975).

[383] Nishihara, K., On a global solution of some quasilinear hyperbolic equation, Tokyo J. Math., 7, (1984), 437–459.

[384] Nunziato, J.W., On heat conduction in materials with memory, Quart. Appl. Math., 29, (1971), 187–204.

[385] Oleinik, O.A., Kruzhkov, S.N., Quasi-linear second-order parabolic equations with many independent variables, Russian Math. Surveys, 16, (1961), 105–146.

[386] Oleinik, O., Some Asymptotic Problems in the Theory of Partial Differential Equations, Cambridge Univ. Press, 1996.

[387] Ordokhani, Y., An application of walsh functions for Fredholm-Hammerstein integro-differential equations, J. Contemp. Math. Sci., 5 (22), (2010), 1055–1063.

[388] Pani, A.K., Fairweather, G., H^1-Galerkin mixed finite element methods for parabolic partial integro-differential equations, IMA J. Numer. Anal., 22, (2002), 231–252.

[389] Pani, A., Thomée, V., Wahlbin, L., Numerical methods for hyperbolic and parabolic integro-differential equations, J. Integral Equations Appl., 4, (1992), 533–584.

[390] Peradze, J., A numerical algorithm for the nonlinear Kirchhoff string equation, Numer. Math., 102, (2005), 311–342.

[391] Peszynska, M., Analysis of an integro-differential equation arising from modelling of flows with fading memory through fissured media, J. Partial Differential Equations, 8, (1995), 159–173.

[392] Peszynska, M., Finite element approximation of diffusion equations with convolution terms, Math. Comp., 65 (215), (1996), 1019–1037.

[393] Petersdorff, T., Schwab, C., Wavelet discretizations of parabolic integrodifferential equations, SIAM J. Numer. Anal., 41, (2003), 159–180.

[394] Pokhožaev, S.I., On a class of quasilinear hyperbolic equations, Mat. Sb., 96, (1975), 152–166 (in Russian). English translation: Math. USSR Sb., 25, (1976), 145–158.

[395] Pokhožaev, S.I., On a quasilinear hyperbolic Kirchhoff equation, Differ. Uravn., 21, (1985), 101–108 (in Russian). English translation: Differ. Equ., 21, (1985), 82–87.

[396] Polyanin, A.D., Manzhirov, A.V. Handbook of Integral Equations. Chapman and Hall/CRC, 2008.

[397] Ponomarev, S.M., On the diffusion of an intense magnetic field into a thin incompressible conductor, Zh. Vychisl. Mat. Mat. Fiz., 27, (1987), 1424–1428 (in Russian). English translation: USSR Comput. Math. Math. Phys., 27 (5), (1987), 98–101.

[398] Prossdorf, S., Silbermann, B., Numerical Analysis for Integral and Related Operator Equations, Akademie-Verlag, Birkhauser Verlag, Berlin, Basel, 1991.

[399] Purcell, E.M., Electricity and Magnetism, Education Development Center, Inc. 1963.

[400] Quarteroni, A., Valli, A., Numerical Approximation of Partial Differential Equations, Springer-Verlag, Berlin, 1994.

[401] Raftari, B., Application of He's homotopy perturbation method and variational iteration method for nonlinear partial integro-differential equations, World Appl. Sci. J., 7 (4), (2009), 399–404.

[402] Rao, S.S., The Finite Element Method in Engineering, 4th Edition, Elsevier, Butterworth Heinemann, 2005.

[403] Reddy, J.N., An Introduction to the Finite Element Method, 3rd Edition, McGraw-Hill, New York, 2005.

[404] Rektorys, K., The Method of Discretization in Time and Partial Differential Equations, Mathematics and Its Applications, Volume 4, (East European Series), D. Reidel, Dordrecht, The Netherlands, 1982.

[405] Renardy, M., Hrusa, W., Nohel, J., Mathematical Problems in Viscoelasticity, Longman Group, Boston, 1987.

[406] Rheinboldt, W.C., Methods for Solving Systems of Nonlinear Equations. SIAM, Philadelphia, 1970.

[407] Richtmeyer, R.D., Morton, K.W., Difference Methods for Initial-Value Problems, Interscience Publishers, New York, 1967.

[408] Rogava, J., Tsiklauri, M., On local convergence of symmetric semi-discrete scheme for abstract analog of Kirchhoff equation, J. Comp. Appl. Math., 236, (2012), 3654–3664.

[409] Russell, D.L., Dirichlet-Neumann boundary control problem associated with Maxwell equations in a cylindrical region, SIAM, J. Math. Anal., 24, (1986), 199–229.

[410] Saadati, R., Raftari, B., Abibi, H., Vaezpour, S.M., Shakeri, S., A comparison between the variational iteration method and trapezoidal rule for solving linear integro-differential equations, World Applied Sciences Journal, 4, (2008), 321–325.

[411] Samarskii, A.A., The Theory of Difference Schemes, Marcel Dekker, Inc., New York, 2001.

[412] Sanikidze, D. G., On the numerical solution of a class of singular integral equations on an infinite interval, Differ. Uravn., 41, (2005), 1280–1285 (in Russian). English translation: Differ. Equ., 41, (2005), 1353–1358.

[413] Sanz-Serna J. M., A numerical method for a partial integro-differential equation, SIAM J. Numer. Anal., 25, (1988), 319–327.

[414] Sattinger, D.H., Monotone methods in nonlinear elliptic and parabolic boundary value problems, Indiana Univ. Math. J., 21, (1972), 979–1000.

[415] Saul'yev, V.K. Integration of Parabolic Equations by the Net Methods, Fizmatgiz, Moscow, 1960 (in Russian). English translation: Integration of Equations of Parabolic Type by the Method of Nets, Int. Series of Monographs in Pure and Appl. Math., Volume 54, Pergamon Press, New York, NY, 1964.

[416] Schatz, A.H., Thomée, V., Wendland, W.L., Mathematical Theory of Finite and Boundary Element Methods, Birkhäuser, Basel-Boston-Berlin, 1990.

[417] Schlesinger, L., Sur les équations intégro-différentielles, C. R. de l'Acad. des Sciences de Paris, 158, (1914), 1872–1875.

[418] Schlesinger, L., Zur Theorie der linearen Integro-differentialgleichungen, Jahresbericht der deutschen Mathematiker-Vereinigung, 24, (1915), 84–123.

[419] Sedov, L.I., Mechanics of Continuous Media, Third ed., Moscow: Nauka, 1976 (in Russian).

[420] Sekar, S., Jaisankar, C., On single term haar wavelet series method for Volterra integro-differential equations, Int. Math. Arch., 5, (2014), 52-58.

[421] Sharma, D., Jiwari, R., Kumar, S., Numerical solution of two point boundary value problems using Galerkin-finite element method, Int. J. Nonlinear Science, 13, (2012), 204–210.

[422] Sharma, D., Jiwari, R., Kumar, S., A comparative study of modal matrix and finite element methods for two point boundary value problems, Int. J. Appl. Math. Mech., 8 (13), (2012), 29–45.

[423] Sharma N., Sharma, K.K., Unconditionally stable numerical method for a nonlinear integro differential equation, Comp. Math. Appl., 67, (2014), 62–76.

[424] Sharma N., Sharma, K.K., Finite element method for a nonlinear parabolic integro-differential equation in higher spatial dimensions, Appl. Math. Modelling (2015), doi: http://dx.doi.org/10.1016/j.apm.2015.02.037.

[425] Shaw, S., Whiteman, J.R., Applications and numerical analysis of partial differential Volterra equations: a brief survey, Comput. Methods Appl. Mech. Engrg., 150, (1997), 397–409.

[426] Shawki, T.G., Clifton, R.J., Shear band formation in thermal viscoplastic materials, Mech. Mater., 8, (1989), 13–43.

[427] Shi, P., Shillor, M., Xiangsheng, X., Existence of a solution to the Stefan problem with Joule's heating, J. Differential Equations, 105 (2), (1993), 239–263.

[428] Showalter, R.E., Monotone Operators in Banach Space and Nonlinear Partial Differential Equations, AMS, Providence, RI, 1997.

[429] Sidi Ammi, M.R., Etude Mathématique de Quelques Problèmes Non Linéaires Issus de la Physique et de L'industrie, Geometry Balkan Press, Bucharest, Romania, 2007 (in French).

[430] Silvestre, L., Hölder continuity for integro-differential parabolic equations with polynomial growth respect to the gradient, Discrete Contin. Dyn. Syst., 28, (2010), 1069–1081.

[431] Simon, L., Application of monotone type operators to parabolic and functional parabolic PDE's, In: C. M. Dafermos, M. Pokorný (eds), Handbook of Differential Equations: Evolutionary Equations, vol 4., North-Holland, Amsterdam, 2008., 267-321.

[432] Sinha, R.K., Ewing, R.E., Lazarov, R.D., Mixed finite element approximations of parabolic integro-differential equations with

nonsmooth initial data, SIAM J. Numer. Anal., 47, (2009), 3269–3292.

[433] Sloan, I.H., Thomée, V., Time discretization of an integro-differential equation of parabolic type, SIAM J. Numer. Anal., 23, (1986), 1052–1061.

[434] Slodicka, M., Busa, J., Jr., Error estimates for the time discretization for nonlinear Maxwell's equations, J. Comput. Math., 26, (2008), 677–688.

[435] Smith, G.D., Numerical Solution of Partial Differential Equations: Finite Difference Methods, Third Edition, Oxford University Press, 1985.

[436] Smyshlyaev, A., Krstic, M., Adaptive Control of Parabolic PDEs, Princeton University Press, 2010.

[437] Smyshlyaev, A., Krstic, M., Closed form boundary state feedbacks for a class of 1-D partial integro-differential equations, IEEE Trans. Automat. Control, 49, (2004), 2185–2202.

[438] Staffans, O.J., On a nonlinear hyperbolic Volterra equation, SIAM J. Math. Anal., 11, (1980), 793–812.

[439] Strang, G., Fix, G.J., An Analysis of the Finite Element Method, Prentice-Hall, Englewood Cliffs, NJ, 1973.

[440] Strikwerda, J.C., Finite Difference Schemes and Partial Differential Equations, Wadsworth & Brooks, Pacific Grove, 1989.

[441] Stupyalis, L., On the spectral problem for the equation of mixed type, Vestnik Leningrad State University Press, 13, (1967), 86–102 (in Russian).

[442] Sun, D., Manoranjan, V.S., Yin, H.-M., Numerical solutions for a coupled parabolic equations arising induction heating processes, Discrete Contin. Dyn. Syst. Supplement, (2007), 956–964.

[443] Tang, T., A finite difference scheme for partial integro-differential equations with a weakly singular kernel, Appl. Numer. Math., 11, (1993), 309–319.

[444] Tavernini, L., Finite difference approximations for a class of semilinear Volterra evolution problems, SIAM J. Numer. Anal., 14 (5), (1977), 931–949.

[445] Teo, K.L., Existence and uniqueness of solutions of system governed by second-order quasilinear integro-partial differential equations of parabolic type, Aequationes Math., 2, (1980), 133–148.

[446] Thomas, J.W., Numerical Partial Differential Equations: Finite Difference Methods, Texts in Applied Mathematics Volume 22, Springer-Verlag, New York, 1995.

[447] Thomée, V., Galerkin Finite Element Methods for Parabolic Problems, Springer-Verlag, New York, NY, 2006.

[448] Thomée, V., Wahlbin, L.B., Long-time numerical solution of a parabolic equation with memory, Math. Comp., 62, (1994), 477–496.

[449] Thomée, V., Zhang, N.Y., Error estimates for semidiscrete finite element methods for parabolic integro-differential equations, Math. Comp., 53, (1989), 121–139.

[450] Thomée, V., Zhang, N.Y., Backward Euler type methods for parabolic integro-differential equations with nonsmooth data, Contributions in numerical mathematics, World Sci. Ser. Appl. Anal., vol. 2, World Sci. Publ., River Edge, NJ, 1993, 373–388.

[451] Tsai, L.Y., On integro-differential equations of parabolic type, Bull. Inst. Math. Acad. Sinica, 9, (1981), 311–320.

[452] Tzavaras, A.E., Shearing of materials exhibiting thermal softening or temperature dependent viscosity, Quart. Appl. Math., 44, (1986), 1–12.

[453] Tzavaras, A.E., Effect of thermal softening in shearing of strain-rate dependent materials, Arch. Ration. Mech. Anal., 99, (1987), 349–374.

[454] Tzavaras, A.E., Non-linear analysis techniques for shear band formation at high strain rates, Appl. Mech. Rev., 45, (1992), 82–94.

[455] Vajnberg, M., Variational Method and Method of Monotone Operators in the Theory of Nonlinear Equations, John Wiley & Sons, Chichester, 1973.

[456] Vasudeva Murthy, A.S., Verwer, J.G., Solving parabolic integro-differential equations by an explicit integration method, J. Comput. Appl. Math., 39, (1992), 121–132.

[457] Vazquez, C., An upwind numerical approach for an American and European option pricing model, Appl. Math. Comput., 97, (1988), 273–286.

[458] Vekua, I.N., On the theory of singular integral equations, Soobshch. Akad. Nauk Gruzin. SSR, 9 (1942), 869–876.

[459] Vekua, I.N., On the integrodifferential equation of Prandtl, Prikl. Mat. Mekh., 9, (1945), 143–150 (in Russian).

[460] Vekua, N.P., Systems of Singular Integral Equations, Gordon & Breach Science Pub, 1967.

[461] Vishik, M.I., On solvability of the boundary value problems for higher order quasilinear parabolic equations, Mat. Sb. (N.S.), 59(101) (supplementary) (1962), 289–325 (in Russian).

[462] Vishik, M.I., Quasi-linear strongly elliptic systems of differential equations of divergence form, Tr. Mosk. Mat. Obs., 12, (1963), 125–184 (in Russian). English translation: Trans. Moscow Math. Soc., 12, (1963), 140–208.

[463] Vlasov, V.V., Rautian, N.A., Shamaev, A.S., Solvability and spectral analysis of integro-differential equations arising in the theory of heat transfer and acoustics, Dokl. Math., 82, (2010), 684–687 (in Russian).

[464] Volk, W., The numerical solution of linear integro-differential equations by projection methods, J. Integral Equ., 9, (1985), 171–190.

[465] Volterra, V., Sopra un problema di elettrostatica, R. Acc. dei Lincei. Trans., 8, Ser. 3, (1884).

[466] Volterra, V., Equazioni integro-differenziali dell' elasticità nel caso della isotropia, R. Acc. dei Lincei. Rend., 18, Ser. 5, 1909.

[467] Volterra, V., Sur les équationes intégro-différentiellee et leurs applications, Acta Math., 35, (1912), 295–356.

[468] Volterra, V., Vibrazioni elastiche nel caso della ereditá, Rendiconti Della Reale Accademia dei Lincei, 21, Ser. 5, 1912, 3–12.

[469] Volterra, V., Theory of Functionals and Integral and of Integral and Integro-Differential Equations, Blackie & Son Ltd, London, 1931.

[470] Wei, W., Yin, H.-M., Global solvability for a singular nonlinear Maxwell's equations, Commun. Pure Appl. Anal., 4, (2005), 431–444.

[471] Woinowsky-Krieger, S., The effect of an axial force on vibration of hinged bars, J. Appl. Mech., 17, (1950), 35–36.

[472] Wollman, S., A deterministic particle method for the Vlasov-Maxwell-Fokker-Planck system in two dimensions, Neural Parallel Sci. Comput., 18, (2010), 461–470.

[473] Wright, T.W., The Physics and Mathematics of Adiabatic Shear Bands, Cambridge University Press, Cambridge, U.K., (2002).

[474] Xing-Ye, Y., Numerical analysis of nonstationary thermistor problem, J. Comput. Math., 12 (3), (1994), 213–223.

[475] Xu, X., A degenerate Stefan-like problem with Joule's heating, SIAM J. Math. Anal., 23, (1992), 1417–1438.

[476] Xu, X., Existence and uniqueness for the nonstationary problem of the electrical heating of a conductor due to the Joule-Thomson effect, Int. J. Math. Math. Sci., 16, (1993), 125–138.

[477] Xu, X., Local and global existence of continuous temperature in the electrical heating of conductors, Houston J. Math., 22, (1996), 435–455.

[478] Yan, Y., Fairweather, G., Orthogonal spline collocation methods for some partial integrodifferential equations, SIAM J. Numer. Anal., 29, (1992), 755–768.

[479] Yanik, E.G., Fairweather, G., Finite element methods for parabolic and hyperbolic partial integro-differential equations, Nonlinear Anal., 12, (1988), 785–809.

[480] Yee, K.S., Numerical solution of initial boundary problems involving Maxwell's equations in isotropic media, IEEE Trans. Antennas and Propagation, 14, (1966), 302–307.

[481] Yildirim, A., Application of He's homotopy perturbation method for solving the Cauchy reaction-diffusion problem, Comput. Math. Appl., 57, (2009), 612–618.

[482] Yildirim, A., Homotopy perturbation method for the mixed Volterra-Fredholm integral equations, Chaos Solitons Fractals, 42, (2009), 2760–2764.

[483] Yin, H.-M., The classical solutions for nonlinear parabolic integrodifferential equations, J. Integral Equations Appl., 1, (1988), 249–263.

[484] Yin, H.-M., Weak and classical solutions of some nonlinear Volterra integrodifferential equations, Comm. Partial Differential Equations, 17, (1992), 1369 – 1385.

[485] Yin, H.-M., Global solutions of Maxwell's equations in an electromagnetic field with a temperature-dependent electrical conductivity, European J. Appl. Math., 5, (1994), 57–64.

[486] Yin, H.-M., Regularity of solutions to Maxwell's system in quasi-stationary electromagnetic fields and applications, Comm. Partial Differential Equations, 22, (1997), 1029–1053.

[487] Yin, H.-M., On Maxwell's equations in an electromagnetic field with the temperature effect, SIAM J. Math. Anal., 29, (1998), 637–651.

[488] Yin, H.-M., On a singular limit problem for nonlinear Maxwell's equations, J. Differential Equations, 156 (2), (1999), 355–375.

[489] Yin, H.-M., On a nonlinear Maxwell's system in quasi-stationary electromagnetic fields, Math. Models Methods Appl. Sci., 14, (2004), 1521–1539.

[490] Yin, H.-M., Existence and regularity of a weak solution to Maxwell's equations with a thermal effect, Math. Methods Appl. Sci., 29 (10), (2006), 1199–1213.

[491] Yin, H.-M., On a Free Boundary Problem Modelling Inductive-Heating Processes, Technical Report 2007-4, Department of Mathematics, Washington State University, Pullman, WA.

[492] Yin, H.-M., On a phase-change problem arising from inductive heating, NoDEA Nonlinear Differential Equations Appl., 13 (5-6), (2007), 735–757.

[493] Young, J.L., Adams, R.S., On the time integration of Maxwell's equations associated with Debye relaxation pro-

cesses, IEEE Trans. Antennas and Propagation, 55, (2007), 2409–2412.

[494] Yuan, G., Regularity of solutions of the thermistor problem, Appl. Anal., 53 (3-4), (1994), 149–156.

[495] Yuzbasi, S., Sahin, N., Sezer, M., Numerical solutions of systems of linear Fredholm integrodifferential equations with Bessel polynomial bases, Comput. Math. Appl., 61, (2011), 3079-3096.

[496] Zang, X.L., Numerical analysis of American option pricing in a jump-diffusion model, Math. Oper. Res., 22, (1997), 668–690.

[497] Zhang, N., On fully discrete Galerkin approximations for partial integro-differential equations of parabolic type, Math. Comp., 60, (1993), 133–166.

[498] Zhao, J., Analysis of finite element approximation for time-dependent Maxwell problems, Math. Comp., 73, (2004), 1089–1105

[499] Zhou, S., Westbrook, D.R., Numerical solutions of the thermistor equations, J. Comput. Appl. Math., 79 (1), (1997), 101–118.

[500] Zhou, Y., Global nonexistence for a quasilinear evolution equation with a generalized Lewis function, Z. Anal. Anwend., 24, (2005), 179–187.

[501] Zhou, Z., Chen, F., Chen, H., Convergence analysis for H^1-Galerkin mixed finite element approximation of one nonlinear integro-differential model, Appl. Math. Comput., 220, (2013), 783–791.

[502] Zhu, Q., Lin, Q., Superconvergence Theory of the Finite Element Methods, Hunan Science Press, 1989.

[503] Zienkiewicz, O.C., The Finite Element Method in Engineering Science, 3rd Edition, McGraw-Hill, London, 1977.

[504] Zlámal, M., Finite element solution of quasistationary nonlinear magnetic field, RAIRO Anal. Numér., 16, (1982), 161–191.

[505] Zolotarevski, V., Zhilin, L., Caraus, I., Approximate solution of singular integro-differential equations over Faber-Laurent polynomials, Differ. Uravn., 40 (2004), 1682–1686 (in Russian). English translation: Differ. Equ., 40, (2004), 1764–1769.

Index

Kolibri 360 GmbH
Ferdinand-Jühlke-Straße 7
99095 Erfurt, Deutschland
produktsicherheit@kolibri360.de